Lecture Notes in Mathematics 2076

Editors:
J.-M. Morel, Cachan
B. Teissier, Paris

For further volumes:
http://www.springer.com/series/304

Valeri Obukhovskii • Pietro Zecca
Nguyen Van Loi • Sergei Kornev

Method of Guiding Functions in Problems of Nonlinear Analysis

Springer

Valeri Obukhovskii
Sergei Kornev
Department of Physics and Mathematics
Voronezh State Pedagogical University
Voronezh, Russia

Pietro Zecca
Dipartimento di Matematica e Informatica
 "U Dini"
Università di Firenze
Firenze, Italy

Nguyen Van Loi
Faculty of Fundamental Science
PetroVietNam University
Ba Ria, Vietnam

ISBN 978-3-642-37069-4   ISBN 978-3-642-37070-0 (eBook)
DOI 10.1007/978-3-642-37070-0
Springer Heidelberg New York Dordrecht London

Lecture Notes in Mathematics ISSN print edition: 0075-8434
                                ISSN electronic edition: 1617-9692

Library of Congress Control Number: 2013937327

Mathematics Subject Classification (2010): 34C25, 34C23, 47J15, 58E07, 34C15, 34B15, 34A60,
                                34G25, 34H05, 47H04, 47H08, 47H09, 47H11, 47J05,
                                49J52, 34K09, 34K13, 34K18, 34K30, 34K35, 54H25,
                                91A23, 93C10, 47N50

Springer is part of Springer Science+Business Media (www.springer.com)

Acknowledgements

The work of V. Obukhovskii and S. Kornev was supported by the Russian FBR Grants 11-01-00328 and 12-01-00392. Professor V. Obukhovskii is much obliged to the University of Florence which supported his work over the book. The work of P. Zecca was supported by a University of Firenze Research Grant.

Contents

1   **Background** 1
   1.1   Multimaps 1
      1.1.1   General Properties 1
      1.1.2   Measurable Multifunctions and Superposition
              Multioperator 5
      1.1.3   Single-Valued Approximations 8
   1.2   Topological Degree 12
   1.3   Coincidence Degree 18
   1.4   Phase Spaces 22
   1.5   Notation 24

2   **Method of Guiding Functions in Finite-Dimensional Spaces** 25
   2.1   Periodic Problem for a Differential Inclusion 25
   2.2   Non-smooth Guiding Functions 36
   2.3   Integral Guiding Functions 39
   2.4   Generalized Periodic Problems 43
      2.4.1   Preliminaries 43
      2.4.2   The Setting of the Problem 44
      2.4.3   Application to Differential Games 45
      2.4.4   Existence Theorem, Corollaries and Example 46
   2.5   Global Bifurcation Problems 50
      2.5.1   Abstract Result 51
      2.5.2   Global Bifurcation of Periodic Solutions 52
      2.5.3   Application 1: Differential Inclusion
              with a Bounded Nonlinearity 63
      2.5.4   Application 2: Global Bifurcation for Functional
              Differential Inclusions 64
      2.5.5   Application 3: Feedback Control System 65

3 Method of Guiding Functions in Hilbert Spaces 69
3.1 Integral Guiding Functions for Differential Inclusions
in Hilbert Spaces 69
3.1.1 The Setting of the Problem 69
3.1.2 Existence of Periodic Solutions 72
3.1.3 Approximation Conditions 77
3.1.4 Application 1: Control Problem of a Partial
Differential Equation 80
3.2 Non-smooth Guiding Functions for Functional
Differential Inclusions with Infinite Delay in Hilbert Spaces 83
3.2.1 Setting of the Problem 83
3.2.2 Existence Theorem 86
3.2.3 Application: Existence of Periodic Solutions
for a Gradient Functional Differential Inclusion 90
3.3 Bifurcation Problem 93
3.3.1 The Setting of the Problem 93
3.3.2 Global Bifurcation Theorem 96
3.3.3 Application 3: Ordinary Feedback Control
Systems in a Hilbert Space 102

4 Second-Order Differential Inclusions 105
4.1 Existence Theorem in an One-Dimensional Space 105
4.2 Applications 110
4.2.1 Equations with Discontinuous Nonlinearities 110
4.2.2 Boundary Value Problem 113
4.2.3 A Second-Order Differential Equation 114
4.2.4 Feedback Control Systems 114
4.2.5 A Model of a Motion of a Particle in a
One-Dimensional Potential 118
4.3 Existence Theorem in Hilbert Spaces 120
4.3.1 Application to a Second-Order Feedback Control
System in Hilbert Space 122
4.3.2 Example 127

5 Nonlinear Fredholm Inclusions and Applications 131
5.1 Preliminaries 131
5.2 Oriented Coincidence Index 133
5.2.1 The Case of a Finite Dimensional Triplet 134
5.2.2 The Case of a Compact Triplet 138
5.2.3 Oriented Coincidence Index for Condensing Triplets 139
5.3 Calculation of the Oriented Coincidence Index by the MGF 145
5.3.1 The Main Result 145
5.3.2 Example 153

5.4   Global Bifurcation Problem 155
      5.4.1   Abstract Result.. 155
      5.4.2   Global Bifurcation for Families of Periodic Trajectories 158
      5.4.3   Example 163

References.. 167

Index.. 175

Introduction

The method of guiding functions (MGF) was originally developed by M.A. Krasnosel'skii and A.I. Perov as one of the tools for solving problems of periodic oscillations in nonlinear systems (see, e.g., [90, 91, 95, 120]). Being geometrically clear and simple to use in applications, it became one of the most powerful and effective instruments for dealing with periodic problems. In the subsequent years it was generalized and extended in various directions. Important aspects of the theory and applications of the MGF were investigated in the works [5, 19–21, 87, 88, 92, 111, 125] and many others. Notice, in particular, that the MGF was extended to differential inclusions and control systems by E.A. Gango and A.I. Povolotskii [63], Yu.G. Borisovich, B.D. Gelman, A.D. Myshkis, and V.V. Obukhovskii [25], and L. Górniewicz and S. Plaskacz (see [64, 67]). In order to study the periodic problem for functional differential equations, A. Fonda [55] introduced the notion of an integral guiding function. The method of integral guiding functions was developed and used in works of S. Kornev and V. Obukhovskii (see [83, 84]). Starting from the works of R.E. Gaines and J.L. Mawhin (see [62, 111]), the notion of a bounding function, closely related to the concept of a guiding function, was systematically used for the study of various boundary value problems by J. Andres, L. Malaguti, V. Taddei, and other researchers (see [5–10, 15]).

In many problems of nonlinear oscillations arises the necessity to use guiding functions which are non-smooth. In particular, such situation appears when different smooth guiding functions are defined in various domains of the phase space of the system. To study these types of problems, F.S. De Blasi, L. Górniewicz, and G. Pianigiani [37] introduced the notion of a non-smooth guiding potential for differential inclusions with convex-valued and non-convex-valued right-hand sides. This notion was extended and developed by G. Gabor and R. Pietkun in [61], S. Kornev and V. Obukhovskii in [82, 84, 85] and M. Filippakis, L. Gasin'ski, and N.S. Papageorgiou in [53] and, by using the methods of non-smooth analysis, applied to various oscillation problems in systems governed by differential inclusions.

It is worth noting that, beginning from the pioneering works, the MGF was applied almost exclusively to objects in finite-dimensional spaces. Only recently, with the use of approximative schemes, the MGF was extended to systems governed

by differential inclusions in infinite-dimensional Hilbert spaces in the papers of
N.V. Loi, V. Obukhovskii, and P. Zecca [100, 108, 109].

Meantime, it was found that the MGF can be not only useful to justify the
existence of oscillations but also successfully applied to the study of the qualitative
behavior of branches of periodic solutions. Using the generalized form of guiding
functions W. Kryszewski considered in [96] the global bifurcation problem for
periodic solutions of first-order differential inclusions in finite-dimensional spaces.
In a cycle of works [101, 102, 104, 107, 109, 115], the systematic investigation and
applications of various modifications of the MGF to the global bifurcation problem
for several types of inclusions (differential and functional differential inclusions,
operator inclusions) were carried out.

Recently, two new branches of the applications of the MGF arose. The first one is
the evaluation of an oriented coincidence index for inclusions containing a nonlinear
Fredholm operator of zero index through the index of guiding functions. In the work
[107] the MGF was used to calculate the oriented coincidence index for a class of
feedback control systems that allowed to obtain the existence result for periodic
trajectories of such systems. The second approach is the application of the MGF
to the study of boundary value problems for second-order differential inclusions
(see [105]).

In our opinion all these directions demonstrate that the MGF plays a remarkable
and important role in problems of contemporary nonlinear analysis. Our target is to
reflect these branches in this monograph.

The plan of the book is as follows.

In order to make the book self-contained, we devote the first chapter to a
detailed description of the fundamental, general properties of multimaps and some
topological characteristics (topological degree and coincidence degree) that will be
used in the next chapters. In particular, we discuss different types of continuity
for multimaps and various operations on multimaps. We describe main properties
of measurable multifunctions and superposition multioperator which is routinely
used when dealing with differential inclusions. We devote particular attention to
the problem of the existence of single-valued approximation for multimaps and
present approximation properties of multimaps. These properties allow to give the
construction and to describe the main features of the topological degree for a wide
class of multimaps. The last part of the first chapter contains the description of
the coincidence degree theory for pairs consisting of zero-index linear Fredholm
operators and multimaps.

In the second chapter, we present the MGF and its modifications for solving
various problems for differential inclusions in finite-dimensional spaces. Starting
from "classical" applications to a periodic problem, we consider non-smooth and
integral guiding functions. We study generalized periodic problems (including
known anti-periodic problem) and consider its applications to differential games.
The last part of the chapter is devoted to applications of the MGF to global
bifurcation problems. After presenting the abstract result, we consider the global
bifurcation of periodic solutions and describe the applications to equations with

discontinuities, ordinary and functional differential inclusions, and feedback control systems.

The third chapter contains the extension of the MGF to the case of differential inclusions in infinite-dimensional Hilbert spaces. To this aim we use the notion of approximate solvability for operator inclusions (this notion is closely related to the notion of A-proper operator developed by F.E. Browder and W.V Petryshyn [30]). Some sufficient conditions for the approximate solvability of inclusions are given. We apply our results to study differential and functional differential inclusions and feedback control systems and to investigate global bifurcation problem for differential inclusions in Hilbert spaces.

In the fourth chapter, by using the MGF, we obtain existence theorems for a boundary value problem for second-order differential inclusions in finite-dimensional and infinite-dimensional Hilbert spaces. It is shown that the abstract result can be applied to study equations with discontinuous nonlinearities, boundary value problems for differential inclusions, and feedback control systems and to the problem of the motion of a particle.

The last chapter is devoted to the nonlinear Fredholm inclusions. After describing the construction of an oriented coincidence index, we present an approach to calculate it through the use of the index of an appropriate guiding function. Furthermore, we prove an abstract global bifurcation theorem for inclusions containing nonlinear Fredholm operators of index zero. We also show how the MGF can be applied to bifurcation problem for feedback control systems with nonlinear Fredholm operators.

Having explained the title and the plan of the book, we would like to stress again that our main goal is to give a self-contained introduction to the method of guiding functions which allows to study effectively various problems arising in the theory of differential inclusions and control systems in finite-dimensional and Hilbert spaces. The book contains all related results of the authors presented in the works [25, 82–86, 99–109, 115, 117].

Chapter 1
Background

1.1 Multimaps

1.1.1 General Properties

In this section we recall some notions of the theory of multivalued maps (details can be found, e.g., in [13, 24, 25, 39, 56, 64, 75, 80] and other sources).

Let X, Y be arbitrary sets and the symbol $P(Y)$ denote the collection of all non-empty subsets of Y.

By *a multivalued map* (or shortly, *multimap*) $\mathscr{F}$ of the set X into the set Y we mean a correspondence which associates to every $x \in X$ a non-empty subset $\mathscr{F}(x) \subseteq Y$, which is called *the value of* x. So, the multimap $\mathscr{F}$ can be written as

$$\mathscr{F} : X \to P(Y).$$

Sometimes we use also the symbols $x \multimap \mathscr{F}(x)$ and $\mathscr{F} : X \multimap Y$.
If $A \subseteq X$, then the set

$$\mathscr{F}(A) = \bigcup_{x \in A} \mathscr{F}(x)$$

is called the *image* of A under $\mathscr{F}$.

The set $\Gamma_{\mathscr{F}} \subseteq X \times Y$, defined by

$$\Gamma_{\mathscr{F}} = \{(x, y) : (x, y) \in X \times Y, \quad y \in \mathscr{F}(x)\}$$

is the *graph* of the multimap $\mathscr{F}$.

For $D \subseteq Y$, the *small pre-image* $\mathscr{F}_+^{-1}(D)$ of the set D is defined by

$$\mathscr{F}_+^{-1}(D) = \{x : x \in X, \ \mathscr{F}(x) \subseteq D\}.$$

V. Obukhovskii et al., *Method of Guiding Functions in Problems of Nonlinear Analysis*,
Lecture Notes in Mathematics 2076, DOI 10.1007/978-3-642-37070-0_1,
© Springer-Verlag Berlin Heidelberg 2013

The *complete pre-image* $\mathscr{F}_-^{-1}(D)$ of the set D is given as

$$\mathscr{F}_-^{-1}(D) = \{x \in X \,:\, \mathscr{F}(x) \cap D \neq \emptyset\}.$$

Now, let X and Y be topological spaces.

Definition 1.1. A multimap $\mathscr{F} : X \to P(Y)$ is *upper semicontinuous* at a point $x \in X$, if, for every open set $V \subset Y$ such that $\mathscr{F}(x) \subset V$, there exists a neighborhood $U(x)$ of x such that $\mathscr{F}(U(x)) \subset V$.

A multimap $\mathscr{F}$ is called u.s.c. if it is upper semicontinuous (u.s.c.) at every point $x \in X$.

The next criterion is obvious.

Proposition 1.1. *The following conditions are equivalent:*

(i) *the multimap $\mathscr{F} : X \to P(Y)$ is u.s.c.;*
(ii) *the set $\mathscr{F}_+^{-1}(V)$ is open for each open $V \subset Y$;*
(iii) *the set $\mathscr{F}_-^{-1}(Q)$ is closed for every closed set $Q \subset Y$.*

Definition 1.2. A multimap $\mathscr{F} : X \to P(Y)$ is called *lower semicontinuous* at a point $x \in X$ if for every open set $V \subseteq Y$ such that $\mathscr{F}(x) \cap V \neq \emptyset$ there exists a neighborhood $U(x)$ of x with the property that $\mathscr{F}(x') \cap V \neq \emptyset$ for all $x' \in V(x)$.

A multimap $\mathscr{F}$ is called lower semicontinuous (l.s.c.) provided it is lower semicontinuous at every point $x \in X$.

The following dual version of Theorem 1.1, presenting some criteria for the lower semicontinuity, holds.

Theorem 1.1. *The following conditions are equivalent:*

(i) *the multimap $\mathscr{F} : X \to P(Y)$ is l.s.c.;*
(ii) *the set $\mathscr{F}_-^{-1}(V)$ is open for every open set $V \subset Y$;*
(iii) *the set $\mathscr{F}_+^{-1}(Q)$ is closed for every closed set $Q \subset Y$.*

Definition 1.3. A multimap $\mathscr{F}$ which is both upper and lower semicontinuous is called *continuous*.

Consider one more important class of multimaps.

Definition 1.4. A multimap $\mathscr{F}$ is called *closed* if its graph $\Gamma_{\mathscr{F}}$ is a closed subset of the space $X \times Y$.

Proposition 1.2. *The following conditions are equivalent:*

(i) *the multimap $\mathscr{F}$ is closed;*
(ii) *for every generalized sequences $\{x_\alpha\} \subset X$, $\{y_\alpha\} \subset Y$, such that $y_\alpha \in \mathscr{F}(x_\alpha)$, if $x_\alpha \to x$ and $y_\alpha \to y$, then $y \in \mathscr{F}(x)$.*

Notice that in the last condition, usual sequences can be used provided X and Y are metric spaces.

Let us introduce some notation:

$$C(Y) = \{D \in P(Y) : D \text{ is closed}\};$$
$$K(Y) = \{D \in P(Y) : D \text{ is compact}\}.$$

If Y is a topological vector space we denote:

$$[Pv(Y) = \{D \in P(Y) : D \text{ is convex}\};$$
$$Cv(Y) = Pv(Y) \cap C(Y) = \{D \in P(Y) : D \text{ is closed and convex}\};$$
$$Kv(Y) = Pv(Y) \cap K(Y) = \{D \in P(Y) : D \text{ is compact and convex}\}.$$

When a multimap F acts into the collections $C(Y)$, $K(Y)$, or $Pv(Y)$, we say that F has closed, compact or convex values respectively.

From the definition it follows immediately that a closed multimap has closed values.

Let Y be a metric space. The function $h : K(Y) \times K(Y) \to \mathbf{R}_+$ defined by

$$h(A, B) = \inf\{\varepsilon > 0 : A \subset V_\varepsilon(B), \ B \subset V_\varepsilon(A)\},$$

where V_ε denotes the ε–neighborhood of a set, is called *the Hausdorff metric* on $K(Y)$.

Proposition 1.3. *Let X be a topological space, Y a metric space. A multimap $\mathscr{F}$: $X \to K(Y)$ is continuous if and only if it is continuous as a single-valued map from X to the metric space $(K(Y), h)$.*

Notice that closed and upper semicontinuous multimaps are a short distance apart. The relation between them is clarified by the following assertions.

Proposition 1.4. *Let X be a topological space, Y a metric space and $\mathscr{F}$: $X \to C(Y)$ a u.s.c. multimap. Then $\mathscr{F}$ is closed.*

To formulate a sufficient condition for a closed multimap to be u.s.c., we need the following definitions.

Definition 1.5. A multimap $\mathscr{F} : X \to P(Y)$ is said to be:

(i) compact if its range $\mathscr{F}(X)$ is relatively compact in Y, i.e., $\overline{\mathscr{F}(X)}$ is compact in Y;

(ii) locally compact if every point $x \in X$ has a neighborhood $U(x)$ such that the restriction of $\mathscr{F}$ to $U(x)$ is compact;

(iii) quasicompact if its restriction to every compact subset $A \subset X$ is compact.

It is clear that $(i) \implies (ii) \implies (iii)$.

Proposition 1.5. *Let $\mathscr{F} : X \to K(Y)$ be a closed locally compact multimap. Then $\mathscr{F}$ is u.s.c.*

The next notion is used in the sequel.

Definition 1.6. Let X be a metric space. A u.s.c. multimap $\mathscr{F} : X \to K(Y)$ which is compact on each bounded subset of X is called completely u.s.c..

Let us mention the following important property of u.s.c. multimaps.

Proposition 1.6. *Let $\mathscr{F} : X \to K(Y)$ be a u.s.c. multimap. If $A \subset X$ is a compact set then its image $\mathscr{F}(A)$ is a compact subset of Y*

The next assertions present continuity properties of some operations on multimaps.

Let X, Y, and Z be topological spaces.

Proposition 1.7. *If multimaps $\mathscr{F}_0 : X \to P(Y)$, and $\mathscr{F}_1 : Y \to P(Z)$ are u.s.c. (l.s.c.) then their composition $\mathscr{F}_1 \circ \mathscr{F}_0 : X \to P(Z)$ defined as*

$$(\mathscr{F}_1 \circ \mathscr{F}_0)(x) = \mathscr{F}_1(\mathscr{F}_0(x)),$$

is u.s.c. (resp. l.s.c.).

Proposition 1.8. *If multimaps $\mathscr{F}_0 : X \to K(Y)$ and $\mathscr{F}_1 : X \to K(Z)$ are u.s.c. (l.s.c.) then their Cartesian product $\mathscr{F}_0 \times \mathscr{F}_1 : X \to K(Y \times Z)$ defined as*

$$(\mathscr{F}_0 \times \mathscr{F}_1)(x) = \mathscr{F}_0(x) \times \mathscr{F}_1(x)$$

is u.s.c. (resp. l.s.c.).

Proposition 1.9. *Let a multimap $F_0 : X \to C(Y)$ be closed, a multimap $F_1 : X \to K(Y)$ u.s.c. and $F_0(x) \cap F_1(x) \neq \emptyset, \forall x \in X$. Then the intersection $F_0 \cap F_1 : X \to K(Y)$, $(F_0 \cap F_1)(x) = F_0(x) \cap F_1(x)$ is u.s.c.*

Now, let X be a topological space, Y a topological vector space.

Proposition 1.10. *If multimaps $\mathscr{F}_0$, $\mathscr{F}_1 : X \to K(Y)$ are u.s.c. (l.s.c.), then their sum $\mathscr{F}_0 + \mathscr{F}_1 : X \to K(Y)$,*

$$(\mathscr{F}_0 + \mathscr{F}_1)(x) = \mathscr{F}_0(x) + \mathscr{F}_1(x)$$

is u.s.c. (resp. l.s.c.).

Proposition 1.11. *If a multimap $\mathscr{F} : X \to K(Y)$ is u.s.c. (l.s.c.), and function $f : X \to \mathbf{R}$ is continuous, then their product $f \cdot \mathscr{F} : X \to K(Y)$,*

$$(f \cdot \mathscr{F})(x) = f(x) \cdot \mathscr{F}(x)$$

is u.s.c. (resp. l.s.c.).

Proposition 1.12. *Let Y be a Banach space. If a multimap $\mathscr{F}: X \to K(Y)$ is u.s.c. (l.s.c.), then its convex closure $\overline{co}\mathscr{F}: X \to Kv(Y)$,*

$$(\overline{co}\mathscr{F})(x) = \overline{co}(\mathscr{F}(x))$$

is u.s.c. (resp. l.s.c.).

1.1.2 Measurable Multifunctions and Superposition Multioperator

In this section we describe the main properties of measurable multifunctions and the superposition multioperator generated by a Carathéodory type multimap. Details can be found in [11, 24, 25, 31, 39, 56, 64, 75, 80] and the references therein.

Let $I \subset \mathbf{R}$ be a compact interval, μ a Lebesgue measure on I and E a Banach space.

Definition 1.7. A multifunction $F : I \to K(E)$ is said to be measurable if for every open subset $V \subset E$ the small pre-image $F_+^{-1}(V)$ is measurable.

It is clear that an equivalent definition is the measurability of the complete pre-image $F_-^{-1}(Q)$ of every closed subset $Q \subset E$. The following assertion gives two more equivalent definitions of the measurability of a multifunction.

Proposition 1.13. *A multifunction $F : I \to K(E)$ is measurable if and only if:*

(i) for every closed set $Q \subset E$ the small pre-image $F_+^{-1}(Q)$ is measurable;
(ii) for every open set $V \subset E$ the complete pre-image $F_-^{-1}(V)$ is measurable.

Notice that from the above definition and Proposition 1.13 it evidently follows that every u.s.c. or l.s.c. multifunction is measurable.

To describe further properties of measurable multifunctions we need the following notions.

Definition 1.8. A function $f : I \to E$ is said to be a measurable selection of a multifunction $F : I \to K(E)$ provided f is measurable and

$$f(t) \in F(t) \quad \text{for } \mu\text{-a.e. } t \in I.$$

The set of all measurable selections of F is denoted as $\mathbf{S}(F)$.

Definition 1.9. A countable family $\{f_n\}_{n=1}^{\infty} \subset \mathbf{S}(F)$ is said to be a Castaing representation of F if

$$\overline{\bigcup_{n=1}^{\infty} f(t)} = F(t)$$

for μ-a.e. $t \in I$.

We say that the multifunction $\tilde{F} : I \to K(E)$ is a step multifunction if there exists a partition of I into a finite family of disjoint measurable subsets $\{I_j\}$, $\cup_j I_j = I$ such that $\tilde{F}$ is constant on each I_j.

Definition 1.10. A multifunction $F : I \to K(E)$ is said to be strongly measurable if there exists a sequence $\{F_n\}_{n=1}^{\infty}$ of step multifunctions such that

$$h(F_n(t), F(t)) \to 0$$

as $n \to \infty$ for μ-a.e. $t \in I$ where h is the Hausdorff metric on $K(E)$.

It is known that in the same manner can be defined the notion of a strongly measurable function and, hence, a strongly measurable selection. Notice that, in general, a measurable multifunction is not a strongly measurable one (see, e.g. [39]). But for compact-valued multifunctions acting into a separable Banach space these notions coincide. It becomes clear from the following assertion describing the main properties of measurable multifunctions (see, e.g. [80]).

Proposition 1.14. *Let E be a separable Banach space. Then for a multimap $F : I \to K(E)$ the following conditions are equivalent:*

(a) F is measurable;
(b) for every countable dense subset $\{x_n\}_{n=1}^{\infty}$ of E the functions $\{\varphi_n\}_{n=1}^{\infty}$, $\varphi_n : I \to \mathbf{R}$,

$$\varphi_n(t) = dist(x_n, F(t))$$

are measurable;
(c) F has a Castaing representation;
(d) F is strongly measurable;
(e) F is measurable as a single-valued map from I into a metric space $(K(E), h)$;
(f) F has a Lusin property: for every $\delta > 0$ there exists a closed subset $I_\delta \subset I$ such that $\mu(I \setminus I_\delta) \le \delta$ and the restriction of F on I_δ is continuous.

If E is an arbitrary (non-separable) space, the following assertion holds (see, e.g. [25]).

Proposition 1.15. *Let E be a Banach space, $F : I \to K(E)$ a strongly measurable multifunction. Then F is measurable and has a Castaing representation consisting of strongly measurable functions.*

Let E be a Banach space and $F : I \to P(E)$ a multifunction. By the symbol $\mathbf{S}^1(F)$ we denote the set of all Bochner integrable selections, i.e.,

$$\mathbf{S}^1(F) = \{f \in L^1(I; E) : f(t) \in F(t) \quad \text{for} \quad \mu - a.e. \, t \in I\}.$$

If $\mathbf{S}^1(F) \ne \emptyset$, then the multifunction F is called *integrable* and its integral is defined as

$$\int_\tau F(s)\,ds = \left\{ \int_\tau f(s)\,ds \; : \; f \in \mathbf{S}^1(F) \right\}$$

for any measurable subset $\tau \subset I$.

It is clear that, if a multifunction $F : I \to K(E)$ is strongly measurable and integrably bounded, i.e., there exists a summable function $v \in L^1_+(I)$ such that

$$\|F(t)\| := \max\{\|y\| : y \in F(t)\} \le v(t) \quad \text{for} \quad \mu - a.e.\,t \in I$$

then F is integrable.

Remark 1.1. Notice that if the multifunction F is constant, $F(t) \equiv A \in Kv(E)$, then $\int_I F(s)\,ds = A\mu(I)$.

Let us consider the following notion. Let E be a Banach space, E_0 a normed space.

Definition 1.11. A multimap $F : I \times E_0 \to K(E)$ is called upper Carathéodory multimap if it satisfies the following conditions:

$(F1)$ for each $x \in E_0$, the multifunction

$$F(\cdot, x) : I \to K(E)$$

admits a strongly measurable selection;

$(F2)$ for a.e. $t \in I$ the multimap $F : E_0 \to K(E)$ is u.s.c.

Remark 1.2. From Proposition 1.14 it follows that in case when the space E is separable, "strongly measurable" in condition $(F1)$ can be replaced with "measurable". In general case Proposition 1.15 implies that, to provide condition $(F1)$, it is sufficient to suppose that the multifunction $F(\cdot, x)$ is strongly measurable for each $x \in E_0$.

The main property of a upper Carathéodory multimap is presented by the following assertion.

Proposition 1.16 (see, e.g. [39, 80]). *If $F : I \times E_0 \to K(E)$ is a upper Carathéodory multimap, then for each strongly measurable function $q : I \to E_0$ there exists a strongly measurable selection $f : I \to E$ of a multifunction $\Phi : I \to K(E)$,*

$$\Phi(t) = F(t, q(t)).$$

Definition 1.12. For a given integer $p \ge 1$, a upper Carathéodory multimap $F : I \times E_0 \to K(E)$ is called L^p-upper Carathéodory if it satisfies additionally the following condition of local integral boundedness:

$(F3)$ for each $r > 0$ there exists a function $v_r \in L^p_+(I)$ such that

$$\|F(t, x)\| := \sup\{\|y\| : y \in F(t, x)\} \leq v_r(t) \quad \text{for } \mu - a.e. \ t \in I$$

for each $x \in E_0$, $\|x\| \leq r$.

Each L^p–upper Carathéodory multimap $F : I \times E_0 \to K(E)$ generates the following *superposition multioperator* $\mathscr{P}_F : C(I; E_0) \to P(L^p(I, E))$,

$$\mathscr{P}_F(x) = \{f \in L^p(I, E) : f(t) \in F(t, x(t)) \text{ for } \mu - a.e. \ t \in I\}.$$

Under additional assumption that the multimap F has convex values, we have the following closedness property of the superposition multioperator (see, e.g. [25, 80])

Proposition 1.17. *Let $F : I \times E_0 \to Kv(E)$ be an L^p–upper Carathéodory multimap, E_1 be a normed space and $A : L^p(I; E) \to E_1$ be a bounded linear operator. Then the composition*

$$A \circ \mathscr{P}_F : C(I; E_0) \longrightarrow Cv(E_1)$$

is a closed multimap.

In practice we commonly use the more popular form of $(F3)$, that is the following sublinear growth:

$(F3')$ there exists a function $\alpha \in L^p_+(I)$ such that

$$\|F(t, x)\| \leq \alpha(t)(1 + \|x\|) \quad \text{for a.e. } t \in I$$

for all $x \in E_0$.

Definition 1.13. For a given integer $p \geq 1$, a multimap $F : I \times E_0 \to K(E)$, satisfying conditions $(F1)$–$(F2)$ and $(F3')$, is called a L^p-upper Carathéodory multimap with α-sublinear growth condition.

Definition 1.14. A multimap $F : \mathbf{R} \times E_0 \to K(E)$ is called T-periodic provided it satisfies the following periodic condition:

(F_T) $F(t, x) = F(t + T, x)$ for $t \in \mathbf{R}$ and all $x \in E_0$.

1.1.3 Single-Valued Approximations

In this section we consider the notion of single-valued approximation of a multimap, important in the fixed point theory and the topological degree theory.

Let (X, ϱ_X), (Y, ϱ_Y) be metric spaces.

Definition 1.15. Let $\mathscr{F} : X \to P(Y)$ be a multimap. For a given $\varepsilon > 0$, a continuous map $f_\varepsilon : X \to Y$ is called an ε-approximation of the multimap $\mathscr{F}$ if for each $x \in X$ there exists $x' \in X$ such that $\varrho_X(x, x') < \varepsilon$ and

$$f_\varepsilon(x) \in V_\varepsilon(\mathscr{F}(x')).$$

for all $x \in X$, where V_ε denotes the ε-neighborhood of a set in the space Y.

It is clear that this notion can be equivalently expressed by the relation

$$f_\varepsilon(x) \in V_\varepsilon(\mathscr{F}(B_\varepsilon(x))) \quad \text{for all} \ x \in X,$$

where $B_\varepsilon(x)$ denotes an open ball in X of radius ε centered at x.

If we introduce the metric ϱ in the Cartesian product $X \times Y$ as

$$\varrho\left((x, y), (x', y')\right) = \max\left\{\varrho_X(x, x'), \varrho_Y(y, y')\right\},$$

then we obtain one more geometrically clear interpretation: the graph Γ_{f_ε} is contained in the ε-neighborhood of the graph $\Gamma_{\mathscr{F}}$.

The following statement on the existence of an ε-approximation holds (see, e.g. [25, 64, 80]).

Proposition 1.18. *Let (X, ϱ) be a metric space, Y a normed space. For each u.s.c. multimap $\mathscr{F} : X \to Cv(Y)$ and $\varepsilon > 0$ there exists a continuous map $f_\varepsilon : X \to Y$ such that*

(*i*) *for every $x \in X$ there exists $x' \in X$ such that $\varrho(x, x') < \varepsilon$ and*

$$f_\varepsilon(x) \cup \mathscr{F}(x) \subset V_\varepsilon(\mathscr{F}(x'));$$

(*ii*)

$$f_\varepsilon(X) \subset co\,\mathscr{F}(X),$$

where co denotes a convex hull of a set.

Definition 1.16. A single-valued ε-approximation satisfying condition (*ii*) of Proposition 1.18 is called regular.

The fact that a map $f : X \to Y$ is an ε–approximation of a multimap $\mathscr{F} : X \to P(Y)$ is written as $f \in a(\mathscr{F}, \varepsilon)$.

Let us mention the following important properties of single-valued approximations, (see, e.g. [25, 64]).

Proposition 1.19. *Let $\mathscr{F} : X \to K(Y)$ be a u.s.c. multimap.*

(*i*) *Let X_1 be a non-empty compact subset of X. Then, for every $\varepsilon > 0$, there exists $\delta > 0$ such that $f \in a(\mathscr{F}, \delta)$ implies $f|_{X_1} \in a(\mathscr{F}_{X_1}, \varepsilon)$.*

(*ii*) *Let X be a compact set, Z a metric space and $\varphi : Y \to Z$ a continuous map. Then, for every $\varepsilon > 0$ there exists $\delta > 0$ such that $f \in a(\mathscr{F}, \delta)$ implies $\varphi \circ f \in a(\varphi \circ \mathscr{F}, \varepsilon)$.*

Our target now is to describe a class of multimaps with non-convex values which also admit single-valued approximations. We need some topological notions.

Definition 1.17. A metric space X is called contractible if there exist a point $x_0 \in X$ and a continuous map (homotopy) $h : X \times [0, 1] \to X$ such that $h(x, 0) = x$ and $h(x, 1) = x_0$ for all $x \in X$.

It is obvious that convex and, more generally, star-shaped sets are contractible.

Definition 1.18. A metric space X is called locally contractible at a point $x_0 \in X$ if for each $\epsilon > 0$ there exists $\delta > 0$, $(\delta \leq \epsilon)$ and a homotopy $h : B_\delta(x_0) \times [0, 1] \to B_\epsilon(x_0)$ such that $h(x, 0) = x$ and $h(x, 1) = x_0$ for all $x \in B_\delta(x_0)$, in other words, the ball $B_\delta(x_0)$ is contractible in $B_\epsilon(x_0)$. A space which is locally contractible at each of its points, is called locally contractible.

It is clear that each union of convex sets in a topological vector space is locally contractible. In particular, each polyhedron is locally contractible.

Definition 1.19 (see [76]). A compact metric space A is called an R_δ-set if there exists a decreasing sequence $\{A_n\}$ of compact contractible sets such that

$$A = \bigcap_{n \geq 1} A_n.$$

Notice that an R_δ-set need not be contractible (see example in [64]). Let us mention also that sets of such topological structure naturally arise in theory of differential equations and inclusions (see below Sect. 2.1).

Definition 1.20 (see [112]). A nonempty compact subset of a metric space X is called aspheric if for each $\epsilon > 0$ there exists $\delta > 0$ $(\delta < \epsilon)$ such that for each $n = 0, 1, \ldots$ every continuous map $g : S^n \to U_\delta(A)$ can be extended to a continuous map $\tilde{g} : \overline{B}^{n+1} \to U_\epsilon(A)$, where $S^n = \{x \in \mathbf{R}^{n+1} : \|x\| = 1\}$, $\overline{B}^{n+1} = \{x \in \mathbf{R}^{n+1} : \|x\| \leq 1\}$.

Definition 1.21. A subset A of a topological space X is called a retract of X if there exists a continuous map (retraction) $r : X \to A$, whose restriction on A is the identity, i.e., $r(x) = x$ for all $x \in A$.

From the Tietze–Dugundji theorem (see, e.g., [29]) it follows that each closed convex subset of a metrizable locally convex topological vector space is the retract of this space.

Definition 1.22. A subset A of a topological space X is called a neighborhood retract provided there exists a retraction $r : U(A) \to A$, where $U(A)$ is a certain neighborhood of A.

Definition 1.23. The embedding of a space X into a space Y is a map $h : X \to Y$ with the following properties:

(i) $h(X) \subset Y$ is a closed set;
(ii) the induced map $\hat{h} : X \to h(X)$ is the homeomorphism.

Definition 1.24. A space X is called an absolute retract (or AR-space) if for each metric space Y and every embedding $h : X \to Y$ the set $h(X)$ is the retract of the space Y. If the set $h(X)$ is the neighborhood retract, then the space X is called the absolute neighborhood retract (or ANR-space).

Notice that the class of ANR-spaces is sufficiently wide. For example, the union of a finite number of closed convex subsets in a normed space is the ANR-space. In particular, each finite polyhedron is the ANR-space.

Moreover, a finite-dimensional compact space is the ANR-space if and only if it is locally contractible (see [29]). In particular, from the Whitney embedding theorem (see, e.g., [74]) it follows that each compact finite dimensional manifold is the ANR-space.

The fact of importance is that for compact subsets of ANR-spaces the notions of aspheric set and R_δ-set coincide:

Proposition 1.20 (see [76]). *If M is a compact subset of an ANR-space, then the following two properties are equivalent:*

(i) *M is an R_δ-set;*
(ii) *M is aspheric.*

Let X, Y be metric spaces. Consider the following class of multimaps.

Definition 1.25. A u.s.c. multimap $\mathscr{F} : X \to K(Y)$ is called J-multimap (or $\mathscr{F} \in J(X, Y)$), provided each set $\mathscr{F}(x)$, $x \in X$ is aspheric.

We have the following characterization of J-multimaps.

Proposition 1.21. *Let Y be an ANR-space. Then a u.s.c. multimap $\mathscr{F} : X \to K(Y)$ is a J-multimap in each of the following cases:*
for each $x \in X$ the set $\mathscr{F}(x)$ is

(i) *a convex set;*
(ii) *an AR-space*
(iii) *a contractible set;*
(iv) *an R_δ-set.*

In particular, it is clear that each continuous single-valued map $f : X \to Y$ is a J-multimap.

Definition 1.26. A multimap $\mathscr{F} : X \to K(Y)$ is called approximable if for every $\varepsilon > 0$ it admits a single-valued ε–approximation and, moreover, for every $\varepsilon > 0$ there exists $\delta_0 > 0$ such that for all δ with $0 < \delta < \delta_0$ and any two δ-approximations $f_\delta, \tilde{f}_\delta : X \to Y$ of the multimap $\mathscr{F}$ there exists a continuous map $h : X \times [0, 1] \to Y$ such that

(i) $h(x, 0) = f_\delta(x)$, $h(x, 1) = \tilde{f}_\delta(x)$ for all $x \in X$;
(ii) $h(\cdot, \lambda) \in a(\mathscr{F}, \varepsilon)$ for each $\lambda \in [0, 1]$.

The main approximation property of a class of J-multimaps can be expressed by the following assertion (see [26, 65, 112]).

Proposition 1.22. *Let X be a compact ANR-space, Y a metric space. Then each J-multimap $\mathcal{F} : X \to K(Y)$ is approximable.*

1.2 Topological Degree

In this section we describe main facts from the topological degree theory for an important class of multivalued maps in a Banach space.

We start with some terminology.

Let $X \subseteq Y$ be certain sets, $\mathcal{F} : X \to P(Y)$ a multimap. A point $x \in X$ is called *a fixed point* of the multimap $\mathcal{F}$ if $x \in \mathcal{F}(x)$. *The set of all fixed points of $\mathcal{F}$ is denoted as Fix$\mathcal{F}$.*

Now we describe the class of multimaps for which we are going to present the topological degree theory. Let X and Y be metric spaces.

Definition 1.27. A multimap $\mathcal{F} : X \to K(Y)$ belongs to the class $CJ(X, Y)$ (or that it is a *CJ*-multimap) if there exist a metric space Z, a J-multimap $\tilde{\mathcal{F}} : X \to K(Z)$, and a continuous map $\varphi : Z \to Y$ such that

$$\mathcal{F} = \varphi \circ \tilde{\mathcal{F}}.$$

The maps $\tilde{\mathcal{F}}$ and φ form the decomposition of $\mathcal{F}$ and we write $\mathcal{F} = (\varphi \circ \tilde{\mathcal{F}})$.

Everywhere in this section E denotes a real Banach space.

Let $X \subseteq E$; each multimap $\mathcal{F} : X \to P(E)$ defines a multimap $\Phi : X \to P(E)$,

$$\Phi(x) = x - \mathcal{F}(x)$$

which is called *the multivalued vector field* or *multifield* corresponding to $\mathcal{F}$. Denoting by $i : X \to E$ the inclusion map, we write

$$\Phi = i - \mathcal{F}.$$

If Λ is a space of parameters, and $\mathcal{G} : X \times \Lambda \to P(E)$ a family of multimaps, then $\Psi : X \times \Lambda \to P(E)$ given as

$$\Psi(x, \lambda) = x - \mathcal{G}(x, \lambda)$$

is called *the family of multifields.*

A point $x \in \Phi(x)$ such that

$$0 \in \Phi(x)$$

is called *a singular point* of a multifield Φ. It is clear that a point x is a singular point of the multifield $\Phi = i - \mathcal{F}$ if and only if it is a fixed point of the multimap $\mathcal{F}$.

In what follows we assume that the reader is familiar with the main facts of the classical Leray–Schauder topological degree theory for compact single-valued maps (see, e.g., [25, 38, 79, 80, 89, 95]).

Let $U \subset E$ be a bounded open subset whose closure we denote as $\overline{U}$ and boundary as ∂U. Let $\mathscr{F} : \overline{U} \to K(E)$ be a compact CJ-multimap such that

$$Fix\mathscr{F} \cap \partial U = \emptyset. \tag{1.1}$$

Our target is to present, following [64], the construction of the topological degree of the multifield $\Phi = i - \mathscr{F}$ and to describe its main properties.

Remark 1.3. Notice that the considered class of CJ-multimaps includes into itself the important collection of (convex compact)-valued u.s.c. multimaps. In fact, in this case the intermediate space Z coincides with E and φ is the identity.

At first, let us suppose that the multimap $\mathscr{F} = (\varphi \circ \tilde{\mathscr{F}})$ is *finite-dimensional*, i.e., there exists a finite-dimensional subspace $E' \subset E$ such that $\mathscr{F}(\overline{U}) \subset E'$. We can assume, w.l.o.g., that the set $U' = U \cap E'$ is non-empty. The closure and the boundary of U' in E' is denoted by $\overline{U'}$ and $\partial U'$ respectively.

It is easy to see that $Fix\mathscr{F}$ is a compact subset of U'.

Lemma 1.1. *There exists $\kappa_0 > 0$ such that, if $\mathscr{O}_\kappa$ is the κ-neighborhood of $Fix\mathscr{F}$ in E' with $0 < \kappa < \kappa_0$, then $\overline{\mathscr{O}}_\kappa \subset U'$ and the multimap $\tilde{\mathscr{F}}$ is approximable on $\overline{\mathscr{O}}_\kappa$.*

Proof. There exists an open neighborhood $\mathscr{N} \subset U'$ of the set $Fix\mathscr{F}$ such that $\overline{\mathscr{N}}$ is the *ANR*-space. Indeed, as such $\mathscr{N}$, we can take a finite union of open balls in E' covering $Fix\mathscr{F}$. Then, from Propositions 1.19(i) and 1.22 it follows that each κ-neighborhood $\mathscr{O}_\kappa$ of $Fix\mathscr{F}$ such that $\overline{\mathscr{O}}_\kappa \subset \mathscr{N}$ is the desirable one. □

The following property can be easily verified.

Lemma 1.2. *If $\mathscr{O}_\kappa$ is the κ-neighborhood of $Fix\mathscr{F}$ satisfying conditions of Lemma 1.1, then there exists $\varepsilon_0 > 0$ such that for each ε-approximation $f_\varepsilon : \overline{\mathscr{O}}_\kappa \to Z$ of the multimap $\tilde{F}$ with $0 < \varepsilon < \varepsilon_0$, the map $\varphi \circ f_\varepsilon : \overline{\mathscr{O}}_\kappa \to E$ is fixed point free ($x \neq \varphi \circ f_\varepsilon(x)$) on the boundary $\partial \mathscr{O}_\kappa$.*

The above lemmas open the possibility to introduce the topological degree of a multifield corresponding to a finite-dimensional CJ-multimap in the following way.

Definition 1.28. Let $\mathscr{F} = (\varphi \circ \tilde{\mathscr{F}}) : \overline{U} \to K(E)$ be a compact finite-dimensional CJ-multimap satisfying (1.1). By a topological degree

$$deg(i - \mathscr{F}, \overline{U})$$

of the corresponding multifield on the set $\overline{U}$ we mean the topological degree of the single-valued field

$$deg(i - \varphi \circ f_\varepsilon, \overline{\mathscr{O}}_\kappa),$$

where $\mathcal{O}_\kappa$ is a neighborhood of $Fix\mathcal{F}$ satisfying conditions of Lemma 1.1 and f_ε is an ε-approximation of $\tilde{\mathcal{F}}$ for $\varepsilon > 0$ sufficiently small.

Using the elementary properties of the Leray–Schauder topological degree, one can easily see that this definition is consistent, i.e., the value $deg(i - \mathcal{F}, \overline{U})$ does not depend on the choice of arbitrary elements of construction: the neighborhood $\mathcal{O}_\kappa$ and the ε-approximation f_ε (if, of course, that κ and ε are sufficiently small). At the same time, it should be mentioned that $deg(i - \mathcal{F}, \overline{U})$ depends on the decomposition $(\varphi \circ \tilde{\mathcal{F}})$: the same multimap $\mathcal{F}$ can admit different decompositions and the degrees defined by these decompositions, in general, can also be different (see examples in [64]).

Now, let us consider the general case when $\mathcal{F} = (\varphi \circ \tilde{\mathcal{F}}) : \overline{U} \to K(E)$ is an arbitrary compact CJ-multimap satisfying condition (1.1).

The following assertion can be easily verified.

Lemma 1.3. *If $\Phi = i - \mathcal{F}$ is the multifield corresponding to the multimap $\mathcal{F}$, then the set $\Phi(\partial U)$ is a closed subset of E.*

Since the set $\Phi(\partial U)$ does not contain zero, the value $\delta_0 = dist(0, \Phi(\partial U))$ is positive. Take the compact set $K = \overline{\mathcal{F}(\overline{U})}$ and choose $0 < \delta < \delta_0$, a finite-dimensional subspace $E' \subset E$ and a continuous map $\pi : K \to E'$ such that

$$\|x - \pi(x)\| < \delta. \tag{1.2}$$

As map π one can take the known Schauder projection (see, e.g., [95]).

Now define the finite-dimensional CJ-multimap $\mathcal{F}' : \overline{U} \to K(E)$ by the decomposition

$$\mathcal{F}' = (\pi\varphi \circ \tilde{\mathcal{F}}).$$

which we call a finite-dimensional approximation of $\mathcal{F} = (\varphi \circ \tilde{\mathcal{F}})$.

Definition 1.29. By a topological degree

$$deg(i - \mathcal{F}, \overline{U})$$

of the multifield corresponding to the multimap $\mathcal{F}$ we mean the topological degree of its finite-dimensional approximation:

$$deg(i - \mathcal{F}', \overline{U}).$$

Again, from the basic properties of the Leray–Schauder degree it follows easily that the value $deg(i - \mathcal{F}, \overline{U})$ does not depend on the particular choice of the finite-dimensional space E' and the projection π.

In the sequel, by the symbol $CJ_{\partial U}(\overline{U}, E)$ we denote the collection of all compact CJ-multimaps $\mathcal{F} : \overline{U} \to K(E)$ satisfying condition (1.1).

We describe now the main properties of the defined characteristics. To present the homotopy invariance property, let us introduce the following notion.

Definition 1.30. Multimaps $\mathcal{F}_0$, $\mathcal{F}_1 \in CJ_{\partial U}(\overline{U}, E)$; $\mathcal{F}_0 = (\varphi_0 \circ \tilde{\mathcal{F}}_0)$, $\mathcal{F}_1 = (\varphi_1 \circ \tilde{\mathcal{F}}_1)$ and the corresponding multifields $\Phi_0 = i - \mathcal{F}_0$, $\Phi_1 = i - \mathcal{F}_1$ are called homotopic

$$\Phi_0 \sim \Phi_1,$$

if there exists a multimap $H \in J(\overline{U} \times [0, 1], Z)$ and a continuous map $k : Z \times [0, 1] \to E$ satisfying the following conditions:

(i) $H(\cdot, 0) = \tilde{\mathcal{F}}_0$, $H(\cdot, 1) = \tilde{\mathcal{F}}_1$;
(ii) $k(\cdot, 0) = \varphi_0$, $k(\cdot, 1) = \varphi_1$;
(iii) the multimap $k \circ H : \overline{U} \times [0, 1] \to K(E)$, defined as

$$(k \circ H)(x, \lambda) = k(H(x, \lambda), \lambda) \quad \text{for all} \quad (x, \lambda) \in \overline{U} \times [0, 1],$$

is compact and fixed point free on $\partial U \times [0, 1]$:

$$x \notin H(x, \lambda), \quad \forall (x, \lambda) \in \partial U \times [0, 1].$$

The multimap $k \circ H$ is called the homotopy in the class $CJ_{\partial U}(\overline{U}, E)$ connecting the multimaps $\mathcal{F}_0$ and $\mathcal{F}_1$ (and corresponding multifields Φ_0 and Φ_1).

(1) *The homotopy invariance property.* Let $\mathcal{F}_0$, $\mathcal{F}_1 \in CJ_{\partial U}(\overline{U}, E)$ and the corresponding multifields $\Phi_0 = i - \mathcal{F}_0$ and $\Phi_1 = i - \mathcal{F}_1$ be homotopic. Then

$$\deg(\Phi_0, \overline{U}) = \deg(\Phi_1, \overline{U}).$$

The idea of the proof consists in passing to a finite-dimensional approximation of the multimap $k \circ H$ and then finding neighborhoods $\mathcal{O}_\kappa$ of the set

$$\{x \in U : x \in H(x, \lambda), \lambda \in [0, 1]\}$$

such that the multimap H is approximable on $\overline{\mathcal{O}}_\kappa \times [0, 1]$. Then the property can be reduced to the homotopy invariance of the topological degree for single-valued fields.

Let us mention in this connection two convenient sufficient conditions for the homotopy of multifields. The first condition shows the stability of the topological degree with respect to "small" perturbations.

Lemma 1.4. *Let $\mathcal{F}_0 \in CJ_{\partial U}(\overline{U}, E)$ and $\tilde{\mathcal{F}} : \overline{U} \to K(Y)$ a CJ-multimap satisfying the boundary condition*

$$\left\| \tilde{\mathcal{F}}(x) \right\| \leq \min \{\|z\| : z \in \Phi_0(x)\}, \forall x \in \partial U,$$

where $\Phi_0 = i - \mathscr{F}_0$. Then, for the multimap $\mathscr{F}_1 = \mathscr{F}_0 + \tilde{\mathscr{F}}$ we have $\mathscr{F}_1 \in CJ_{\partial U}(\overline{U}, E)$ and

$$\Phi_1 = i - \mathscr{F}_1 \sim \Phi_0$$

implying $deg(\Phi_0, \overline{U}) = deg(\Phi_1, \overline{U})$.

The second condition shows the homotopy property of multifields which do not allow opposite directions on ∂U.

Lemma 1.5. *Let the multimaps $\mathscr{F}_0, \mathscr{F}_1 \in CJ_{\partial U}(\overline{U}, E)$ satisfy the boundary condition*

$$\frac{z_0}{\|z_0\|} \neq -\frac{z_1}{\|z_1\|}$$

for all $z_0 \in \Phi_0(x)$, $z_1 \in \Phi_1(x)$, $x \in \partial U$, where $\Phi_k = i - \mathscr{F}_k$, $k = 0, 1$. Then, $\Phi_0 \sim \Phi_1$ and hence

$$deg(\Phi_0, \overline{U}) = deg(\Phi_1, \overline{U}) .$$

(2) *Additive dependence on the domain.* Let $\{U_j\}_{j=1}^{m}$ be a family of open disjoint subsets of U and the compact CJ-multimap $\mathscr{F} : \overline{U} \to K(Y)$ has no fixed points on the set $\overline{U} \setminus \bigcup_{j=1}^{m} U_j$. Then

$$\deg(i - \mathscr{F}, \overline{U}) = \sum_{j=1}^{m} \deg(i - \mathscr{F}, \overline{U}_j).$$

The idea of the proof is to pass to a finite-dimensional approximation and then to use Proposition 1.19. After that, apply the corresponding property for single-valued fields.

The following useful property also can be verified by the passing to approximations and the application of the corresponding "single-valued" property.

(3) *The principle of map restriction.* Let $E_1 \subset E$ be a closed subspace and a multimap $\mathscr{F} \in CJ_{\partial U}(\overline{U}, E)$ is such that $\mathscr{F}(\overline{U}) \subset E_1$. Then

$$\deg(i - \mathscr{F}, \overline{U}) = \deg_{E_1}(i - \mathscr{F}, \overline{U}_1),$$

where $U_1 = U \cap E_1$ and $\deg_{E_1}$ means the degree evaluated in the space E_1.

(4) *The fixed point principle.* Let $\mathscr{F} \in CJ_{\partial U}(\overline{U}, E)$ and

$$\deg(i - \mathscr{F}, \overline{U}) \neq 0.$$

Then $\emptyset \neq Fix\mathscr{F} \subset U$.

Idea of the proof: If $\mathscr{F}$ has no fixed points on the whole $\overline{U}$, then, subsequently, we can construct a finite-dimensional approximation $\mathscr{F}'$ and a single-valued approximation $\varphi \circ f_\varepsilon$ which also are fixed point free. But then the degree of the field $i - \varphi \circ f_\varepsilon$ on the corresponding domain equals zero, and hence $\deg(i - \mathscr{F}, \overline{U}) = 0$, contrary to the assumption.

This general principle allows a simple proof of many fixed point theorems used in applications. Let us present the following generalization of the well-known Bohnenblust–Karlin theorem (see [22]).

Proposition 1.23. *Let M be a non-empty closed convex subset of E and $\mathscr{F} : M \to K(M)$ be a compact CJ– multimap. Then $Fix\mathscr{F} \neq \emptyset$.*

Proof. Let $\mathscr{F}$ have the decomposition $\mathscr{F} = (\varphi \circ \tilde{\mathscr{F}})$. Take an arbitrary open bounded convex set $U \subset E$ containing the compact convex set $M_1 = \overline{co}\mathscr{F}(M)$. Consider now the retraction $\rho : \overline{U} \to M_1$ and the multimap $\mathscr{F}_1 = \mathscr{F} \circ \rho : \overline{U} \multimap M_1$. It is clear that $\mathscr{F}_1$ is the compact CJ-multimap with the decomposition $\mathscr{F}_1 = (\varphi \circ \tilde{\mathscr{F}} \circ \rho)$ and its fixed points coincide with fixed points of $\mathscr{F}$.

Take any point $x_0 \in U$ and consider the deformation $k : Z \times [0, 1] \to E$ given as

$$k(z, \lambda) = (1 - \lambda)\varphi(z) + \lambda x_0$$

and the multimap $H : \overline{U} \times [0, 1] \to K(Z)$,

$$H(x, \lambda) = \tilde{\mathscr{F}} \circ \rho(x), \quad \forall \lambda \in [0, 1]$$

Then $k \circ H$ defines the homotopy joining the multimap $\mathscr{F}_1$ with the constant map $H(\cdot, 1) \equiv x_0$. By using the normalization property of the "single-valued" degree and the homotopy invariance property (1) we obtain

$$\deg(i - \mathscr{F}_1, \overline{U}) = \deg(i - H(\cdot, 1), \overline{U}) = 1,$$

from where it follows that $Fix\mathscr{F}_1 = Fix\mathscr{F}$ is a non-empty set. $\square$

By following similar homotopy methods, the next *principle of forbidden direction* can be verified.

Proposition 1.24. *Let $\mathscr{F} : \overline{U} \to K(E)$ be a compact CJ-multimap, $\Phi = i - \mathscr{F}$ the corresponding multifield. Suppose that for a given point $a \in U$ we have*

$$\Phi(x) \cap L_x^a = \emptyset, \quad \text{for each } x \in \partial U,$$

where

$$L_x^a = \{y \in E : y = \mu(x - a), \ \mu \leq 0\}.$$

Then $\deg(\Phi, \overline{U}) = 1$ and hence $\emptyset \neq Fix\mathscr{F} \subset U$.

As a corollary we can obtain the following fixed point principle for multimaps satisfying *the Leray–Schauder boundary condition.*

Proposition 1.25. *Let U be a bounded open neighborhood of zero and a compact CJ-multimap $\mathscr{F} : \overline{U} \to K(E)$ satisfies the boundary condition*

$$x \notin \lambda \mathscr{F}(x)$$

for all $x \in \partial U$ and $0 < \lambda \leq 1$. Then $\deg(\Phi, \overline{U}) = 1$ and hence $\emptyset \neq Fix\mathscr{F} \subset U$.

In conclusion, let us present a version of *the odd field theorem* that can be proved by the same approximation and homotopy methods.

Proposition 1.26. *Let U be a symmetric bounded neighborhood of zero, $\mathscr{F} : \overline{U} \to K(E)$ be a compact CJ-multimap such that the corresponding multifield $\Phi = i - \mathscr{F}$ satisfies the following boundary condition:*

$$\Phi(x) \cap \mu\Phi(-x) = \emptyset \quad \text{for all } x \in \partial U, \ 0 \leq \mu \leq 1.$$

Then

$$\deg(\Phi, \overline{U}) \equiv 1 \ (mod \ 2)$$

and hence $\emptyset \neq Fix\mathscr{F} \subset U$.

1.3   Coincidence Degree

In this section we consider the coincidence degree for a pair consisting of a linear Fredholm operator of zero index and a multimap.

Let E_1 and E_2 be Banach spaces; $L : Dom \, L \subseteq E_1 \to E_2$ a linear operator.

Recall the following known facts (see, e.g., [62, 111]).

Proposition 1.27. *Let $P : E_1 \to E_1$ be a linear projection operator such that $Im \, P = Ker \, L$. Then*

(i) the operator $L_P : Dom \, L \cap Ker \, P \to Im \, L$ given as

$$L_P(x) = L(x) \quad \text{for all } x \in Dom \, L \cap Ker \, P$$

   is a linear isomorphism;

(ii) the operator $K_P : Im \, L \to Dom \, L \cap Ker \, P$ given as

$$K_P = L_P^{-1},$$

   satisfies the relation

$$K_P \circ Lx = x - Px \quad \text{for all } x \in Dom \, L.$$

Definition 1.31. A linear operator $L : Dom\,L \to E_2$ is called a linear Fredholm operator of zero index if the spaces $Ker\,L$ and $Coker\,L = E_2/Im\,L$ have a finite dimension and

$$dim\,Ker\,L = dim\,Coker\,L.$$

Proposition 1.28. *Let $L : Dom\,L \to E_2$ be a linear Fredholm operator of zero index such that $Im\,L$ is a closed subspace of E_2. Then*

(*i*) *there exist linear continuous projection operators $P : E_1 \to E_1$ and $Q : E_2 \to E_2$ such that $Im\,P = Ker\,L$ and $Im\,L = Ker\,Q$;*
(*ii*) *the canonical projection $\Pi : E_2 \to E_2/Im\,L$, given as*

$$\Pi y = y + Im\,L,$$

is a continuous linear operator;
(*iii*) *there exists a continuous linear isomorphism $\Lambda : Coker\,L \to Ker\,L$;*
(*iv*) *the equation*

$$Lx = y, \quad y \in E_2$$

is equivalent to the equation

$$(i - P)x = (\Lambda\Pi + K_{P,Q})(y),$$

where i is the identity in E_1 and the operator $K_{P,Q} : E_2 \to E_1$ is given by the relation:

$$K_{P,Q}(y) = K_P(y - Qy).$$

Our target now is to extend the topological degree theory developed in the previous section, by describing the coincidence degree theory for linear Fredholm multimaps of zero index and *CJ*-multimaps.

Let $U \subset E_1$ be an open bounded set.

Definition 1.32. A *CJ*-multimap $\mathscr{F} : \overline{U} \to K(E_2)$ is called *L*-compact if the composition

$$(\Lambda\Pi + K_{P,Q}) \circ \mathscr{F} : \overline{U} \to K(E_1)$$

is a compact multimap.

Remark 1.4. The definition of an *L*-compact multimap does not depend on the choice of linear projection operators $P : E_1 \to E_1$, $Q : E_2 \to E_2$, and the isomorphism $\phi : Coker\,L \to Ker\,L$.

Definition 1.33. A point $x \in Dom\,L$ is called a coincidence point of the operator L and the multimap $\mathscr{F}$ if

$$Lx \in \mathscr{F}(x).$$

The set of all coincidence points of L and $\mathscr{F}$ is denoted by $Coin\,(L, \mathscr{F})$.

Consider now the multimap $\mathscr{G} : \overline{U} \to K(E_1)$ of the form

$$\mathscr{G}(x) = Px + (\Lambda\Pi + K_{P,Q}) \circ \mathscr{F}(x), \quad x \in \overline{U}. \tag{1.3}$$

From Proposition 1.28 (iv) it follows that $Fix\,\mathscr{G}$ coincides with $Coin\,(L, \mathscr{F})$.

Lemma 1.6. *The multimap $\mathscr{G}$ defined by (1.3) is a compact CJ-multimap.*

Proof. In fact, let $\mathscr{F} = (\varphi \circ \tilde{\mathscr{F}})$; $\tilde{\mathscr{F}} : \overline{U} \to K(Z), \varphi : Z \to E_1$ be the decomposition of $\mathscr{F}$. It is clear that the multimap $\hat{F} : \overline{U} \to K(E_1 \times Z)$ defined as

$$\hat{F}(x) = \{x\} \times \tilde{F}(x)$$

is a J-multimap.

Further, define the map $\Psi : E_1 \times Z \to E_1$ as

$$\Psi(x, z) = Px + (\Lambda\Pi + K_{P,Q})(z).$$

Then $\mathscr{G}$ can be decomposed as $(\Psi \circ \hat{F})$. The compactness of $\mathscr{G}$ follows from the L-compactness of $\mathscr{F}$ and the fact that the projection P has the finite-dimensional range. □

Denote by $CJ^L(\overline{U}, E_2)$ the class of all L-compact CJ-multimaps $\mathscr{F} : \overline{U} \to K(E_2)$. The sub-class of $CJ^L(\overline{U}, E_2)$ consisting of all such multimaps $\mathscr{F}$ for which

$$Coin\,(L, \mathscr{F}) \bigcap (\partial U \cap Dom\,L) = \emptyset$$

is denoted by $CJ^L_{\partial U}(\overline{U}, E_2)$.

Definition 1.34. By the coincidence degree

$$\deg(L, \mathscr{F}, \overline{U})$$

of a pair $(L, \mathscr{F})$, where $\mathscr{F} \in CJ^L_{\partial U}(\overline{U}, E_2)$ we mean the topological degree $\deg(\Phi, \overline{U})$ of the multifield $\Phi = i - \mathscr{G}$ corresponding to the multimap $\mathscr{G} : \overline{U} \to K(E_1)$ given by (1.3).

From the definition it follows that the introduced characteristic possesses the main properties of the topological degree. Let us list them.

Definition 1.35. Multimaps $\mathscr{F}_0, \mathscr{F}_1 \in CJ^L_{\partial U}(\overline{U}, E_2)$; $\mathscr{F}_0 = (\varphi_0 \circ \tilde{\mathscr{F}}_0)$, $\mathscr{F}_1 = (\varphi_1 \circ \tilde{\mathscr{F}}_1)$ are called L-homotopic

$$\mathscr{F}_0 \overset{L}{\sim} \mathscr{F}_1,$$

if there exists a multimap $H \in J(\overline{U} \times [0,1], Z)$ and a continuous map $k : Z \times [0,1] \to E_2$ satisfying the following conditions:

(i) $H(\cdot, 0) = \tilde{\mathscr{F}}_0$, $H(\cdot, 1) = \tilde{\mathscr{F}}_1$;

(ii) $k(\cdot, 0) = \varphi_0$, $k(\cdot, 1) = \varphi_1$;

(iii) the multimap $k \circ H : \overline{U} \times [0,1] \to K(E_2)$, defined as

$$(k \circ H)(x, \lambda) = k(H(x, \lambda), \lambda) \quad \text{for all } (x, \lambda) \in \overline{U} \times [0,1],$$

is L-compact (in the sense of Definition 1.32) and coincidence point free on $(\partial U \cap Dom\, l) \times [0,1]$:

$$Lx \notin k(H(x, \lambda), \lambda) \quad \forall (x, \lambda) \in (\partial U \cap Dom\, L) \times [0,1].$$

Proposition 1.29. *If* $\mathscr{F}_0 \overset{L}{\sim} \mathscr{F}_1$, *then*

$$\deg(L, \mathscr{F}_0, \overline{U}) = \deg(L, \mathscr{F}_1, \overline{U}).$$

Proposition 1.30. *Let* $\{U_j\}_{j=1}^{m}$ *be a family of open disjoint subsets of* U *and* L-*compact multimap* $\mathscr{F} \in CJ(\overline{U}, E_2)$ *is such that*

$$Coin(L, \mathscr{F}) \bigcap \left((\overline{U} \setminus \bigcup_{j=1}^{m} U_j) \cap Dom\, L \right) = \emptyset.$$

Then

$$\deg(L, \mathscr{F}, \overline{U}) = \sum_{j=1}^{m} \deg(L, \mathscr{F}, \overline{U}_j).$$

Proposition 1.31. *Let* $\mathscr{F} \in CJ_{\partial U}^{L}(\overline{U}, E_2)$ *and*

$$\deg(L, \mathscr{F}, \overline{U}) \neq 0.$$

Then $\emptyset \neq Coin(L, \mathscr{F}, \overline{U}) \subset (U \cap Dom\, L)$.

As an application of this general principle, let us consider the following assertion (see [39]) which we need in the sequel.

Theorem 1.2. *Let* $\mathscr{F} : \overline{U} \to K(E_2)$ *be a CJ-multimap such that the multimaps* $\Pi\mathscr{F}$ *and* $K_{P,Q}\mathscr{F}$ *are compact and the following conditions are fulfilled:*

(i) $Lx \notin \lambda\mathscr{F}(x)$ *for all* $\lambda \in (0,1]$, $x \in Dom\, L \cap \partial U$;

(ii) $0 \notin \Pi\mathscr{F}(x)$ *for all* $x \in Ker\, L \cap \partial U$;

(iii)

$$\deg_{Ker\, L}(\Lambda\Pi\mathscr{F}|_{\overline{U}_{Ker\, L}}, \overline{U}_{Ker\, L}) \neq 0,$$

where the symbol $\deg_{Ker\,L}$ *means the topological degree evaluated in the space* $Ker\,L$, *and* $\overline{U}_{Ker\,L} = \overline{U} \cap Ker\,L$.

Then $\emptyset \neq Coin(L,\mathscr{F},\overline{U}) \subset (U \cap Dom\,L)$.

Proof. Consider the deformation $\Psi : \overline{U} \times [0,1] \to K(E_1)$, given as

$$\Psi(x,\lambda) = Px + (\Lambda\Pi + \lambda K_{P,Q})\mathscr{F}(x), \quad (x,\lambda) \in \overline{U} \times [0,1].$$

For $\lambda \in (0,1]$ and $x \in Dom\,L \cap \partial U$, taking into account that $\frac{1}{\lambda}\Lambda$ is also a linear isomorphism of the spaces $Coker\,L$ and $Ker\,L$, we come to the conclusion that from condition (i) it follows that

$$x \notin \Psi(x,\lambda).$$

On the other side, condition (ii) implies that

$$x \notin \Psi(x,0)$$

for all $x \in Dom\,L \cap \partial U$.

So, Ψ generates an homotopy of CJ-multimaps, implying

$$\deg(L,\mathscr{F},\overline{U}) = \deg(i - P - \Lambda\Pi\mathscr{F},\overline{U}),$$

where in the right-hand side we have the topological degree of the multifield corresponding to the finite-dimensional multimap $P + \Lambda\Pi\mathscr{F}$. By applying the principle of map restriction, we obtain

$$\deg(i - P - \Lambda\Pi\mathscr{F},\overline{U}) = \deg_{Ker\,L}(-\Lambda\Pi\mathscr{F}|_{\overline{U}_{Ker\,L}}, \overline{U}_{Ker\,L}).$$

It remains only to apply condition (iii) and Proposition 1.31. Notice that constructions of a coincidence degree for pairs consisting of a linear Fredholm operators of zero index and convex-valued multimaps were suggested in [123, 128]. The coincidence degree for pairs with a nonlinear Fredholm operator will be presented in Chapter 5 below.   □

1.4 Phase Spaces

In considering functional differential inclusions with infinite delay we use an axiomatic definition of the *phase space* $\mathscr{B}$, introduced by J.K.Hale and J.Kato (see [71,73]). In this section we describe briefly its main properties.

The space $\mathscr{B}$ is the linear topological space of functions mapping $(-\infty,0]$ into a Hilbert space H endowed with a seminorm $\|\cdot\|_{\mathscr{B}}$.

For $T > 0$ and any function $y : (-\infty;T] \to H$ and for every $t \in I = [0,T]$, y_t represents the function from $(-\infty,0]$ into H defined by

$$y_t(\theta) = y(t + \theta), \ \theta \in (-\infty; 0].$$

We assume that $\mathscr{B}$ satisfies the following axioms.

($\mathscr{B}1$) If $y : (-\infty; T] \to H$ is such that $y_{|_I} \in C(I; H)$ and $y_0 \in \mathscr{B}$, then we have

(i) $y_t \in \mathscr{B}$ for $t \in I$;

(ii) the function $t \in I \mapsto y_t \in \mathscr{B}$ is continuous;

(iii) $\|y_t\|_{\mathscr{B}} \leq K(t) \sup\limits_{0 \leq \tau \leq t} \|y(\tau)\| + N(t)\|y_0\|_{\mathscr{B}}$ for $t \in [0, T]$, where $K(\cdot), N(\cdot) :$
$[0; \infty) \to [0; \infty)$ are independent of y, $K(\cdot)$ is strictly positive and continuous, and $N(\cdot)$ is bounded.

($\mathscr{B}2$) There exists $l > 0$ such that

$$\|\psi(0)\|_H \leq l\|\psi\|_{\mathscr{B}}$$

for all $\psi \in \mathscr{B}$.

Let us mention that under the above hypotheses the space C_{00} of all continuous functions from $(-\infty, 0]$ into H with compact support is a subset of each phase space $\mathscr{B}$ ([73], Proposition 1.2.1). We assume, additionally, that the following hypothesis holds.

($\mathscr{B}3$) If a uniformly bounded sequence $\{\psi_n\}_{n=1}^{+\infty} \subset C_{00}$ converges to a function ψ compactly (i.e. uniformly on each compact subset of $(-\infty, 0]$), then $\psi \in \mathscr{B}$ and

$$\lim_{n \to +\infty} \|\psi_n - \psi\|_{\mathscr{B}} = 0.$$

The hypothesis ($\mathscr{B}3$) states that the Banach space $BC((-\infty, 0]; H)$ of bounded continuous functions is continuously embedded into $\mathscr{B}$.

We can consider the following examples of phase spaces satisfying all above properties.

(1) For $\nu > 0$, let $\mathscr{B} = C_\nu$ be the space of functions $\psi : (-\infty; 0] \to H$ such that: (i) $\psi_{|[-r,0]} \in C([-r, 0]; E)$ for each $r > 0$; (ii) the limit $\lim\limits_{\theta \to -\infty} e^{\nu\theta}\|\psi(\theta)\|$ is finite. Then we set

$$\|\psi\|_{\mathscr{B}} = \sup_{-\infty < \theta \leq 0} e^{\nu\theta}\|\psi(\theta)\|.$$

(2) *Spaces of "fading memory".* Let $\mathscr{B} = C_\rho$ be the space of functions $\psi : (-\infty; 0] \to E$ such that

(a) $\psi \in C([-r; 0]; E)$ for some $r > 0$;

(b) ψ is Lebesgue measurable on $(-\infty; -r)$ and there exists a positive Lebesgue integrable function $\rho : (-\infty; -r) \to R^+$ such that $\rho\psi$ is Lebesgue integrable on $(-\infty; -r)$; moreover, there exists a locally bounded function $P : (-\infty; 0] \to R^+$ such that, for all $\xi \leq 0$, $\rho(\xi + \theta) \leq P(\xi)\rho(\theta)$ a.e. $\theta \in (-\infty; -r)$. Then,

$$\|\psi\|_{\mathscr{B}} = \sup_{-r \leq \theta \leq 0} \|\psi(\theta)\| + \int_{-\infty}^{-r} \rho(\theta)\|\psi(\theta)\|d\theta.$$

A simple example of such a space can be obtained by taking the function $\rho(\theta) = e^{\mu\theta}$, $\mu \in R$.

1.5   Notation

Let E be a Banach space; $k, p \geq 1$ be integer numbers. In this monograph we use the following notation:

By $B_E(0, r)$—a closed ball of radius r in E centered at 0.

$C([0, T]; E)$—the space of all continuous functions $x: [0, T] \to E$ with the norm

$$\|x\|_C = \sup_{t \in [0,T]} \|x(t)\|.$$

$L^p([0, T]; E)$—the space of all p-summable functions with the norm:

$$\|x\|_p = \left(\int_0^T \|x(t)\|^p dt \right)^{\frac{1}{p}}.$$

$W^{k,p}([0, T]; E)$—the Sobolev space of functions with the norm:

$$\|x\|_W = \left(\|x\|_p^p + \|x'\|_p^p + \cdots + \|x^{(k)}\|_p^p \right)^{\frac{1}{p}}.$$

By $C_T([0, T]; E)$ $[W_T^{k,p}([0, T]; E)]$ we denote the subspaces consisting of all functions of $C([0, T]; E)$ [resp., $W^{k,p}([0, T]; E)$] satisfying the periodic boundary condition:

$$x(0) = x(T).$$

By $B_C(0, r)$ $[B_{C_T}(0, r)]$ we denote the balls of radius r in $C([0, T]; E)$ [resp., in $C_T([0, T]; E)$] centered at 0.

The symbols $\langle \cdot, \cdot \rangle$ $[|\cdot|]$ denote the inner product [resp., norm] in $\mathbf{R}^n$.

For a Hilbert space H by $\langle \cdot, \cdot \rangle_H$ we denote the inner product in H.

Chapter 2
Method of Guiding Functions
in Finite-Dimensional Spaces

2.1 Periodic Problem for a Differential Inclusion

In this section we present the guiding functions method for studying the periodic problem for a differential inclusion in a finite-dimensional space.

We start considering a differential inclusion in a finite-dimensional space $\mathbf{R}^n$ of the following form:

$$x'(t) \in F(t, x(t)), \quad a.e. \quad t \in [0, T] \tag{2.1}$$

where $F : [0, T] \times \mathbf{R}^n \to Kv(\mathbf{R}^n)$ is an L^1-upper Carathéodory multimap.

By a *solution* of inclusion (2.1) we mean an absolutely continuous function $x : [0, T] \to \mathbf{R}^n$ satisfying (2.1) for a.e. $t \in [0, T]$. It is well known (see, e.g., [25, 80]) that the L^1-upper Carathéodory condition implies the existence of *a local solution* to the Cauchy problem corresponding to (2.1), i.e., a solution defined on some interval $[0, h]$, $0 < h \leq T$ and satisfying the initial condition

$$x(0) = x_0 \in \mathbf{R}^n. \tag{2.2}$$

To guarantee the existence of a *global solution* ($h = T$) it is sufficient to strengthen the condition posed on the multimap F supposing that F is an L^1-upper Carathéordory multimap with α-sublinear growth. More precisely, the following assertion holds (see, e.g., [24, 39, 64, 80])

Proposition 2.1. *If* $F : [0, T] \times \mathbf{R}^n \to Kv(\mathbf{R}^n)$ *is an* L^1-*upper Carathéodory multimap with* α-*sublinear growth, then for each* $x_0 \in \mathbf{R}^n$, *the solution set* $\Pi_F(x_0)$ *of the Cauchy problem*

$$x'(t) \in F(t, x(t)) \quad a.e. \quad t \in [0, T] \tag{2.3}$$

$$x(0) = x_0 \tag{2.4}$$

V. Obukhovskii et al., *Method of Guiding Functions in Problems of Nonlinear Analysis*,   25
Lecture Notes in Mathematics 2076, DOI 10.1007/978-3-642-37070-0_2,
© Springer-Verlag Berlin Heidelberg 2013

is an R_δ-set in the space $C([0, T]; \mathbf{R}^n)$ endowed with the usual norm of uniform convergence.

Moreover, the multimap $\Pi_F : \mathbf{R}^n \to K(C([0, T]; \mathbf{R}^n))$, $x \multimap \Pi_F(x)$ is u.s.c.

Now, we say that a solution x to differential inclusion (2.1) is T-*periodic* if it satisfies the following boundary value condition of periodicity

$$x(0) = x(T). \tag{2.5}$$

It is clear that such function can be extended to a T-periodic solution defined on $\mathbf{R}$ provided that F is T-periodic, i.e., the multimap $F : \mathbf{R} \times \mathbf{R}^n \to Kv(\mathbf{R}^n)$ satisfies $F(t + T, \cdot) = F(t, \cdot)$ for all $t \in [0, T]$.

In order to study periodic problem (2.1), (2.5) we can introduce the *translation multioperator* along the trajectories of (2.1), (2.5) in the following way. For any $t \in [0, T]$ let $\theta_t : C([0, T]; \mathbf{R}^n) \to \mathbf{R}^n$ be the *evaluation map* defined as

$$\theta_t(y) = y(t).$$

Then the multioperator $\mathbf{P}_t^F : \mathbf{R}^n \multimap \mathbf{R}^n$ given as the composition

$$\mathbf{P}_t^F(x) = \theta_t \circ \Pi_F(x),$$

is called the translation multioperator along the trajectories of problem (2.1), (2.5), or simply, the translation multioperator.

The following assertion is evident

Proposition 2.2. *Periodic problem (2.1), (2.5) has a solution if and only if the corresponding translation multioperator $\mathbf{P}_T^F$ has a fixed point $x_* \in \mathbf{R}^n$, $x_* \in \mathbf{P}_T^F(x_*)$.*

As a direct consequence of Propositions 2.1 and 2.2 we get the following general existence result for problem (2.1), (2.5).

Theorem 2.1. *Let $U \subset \mathbf{R}^n$ be an open bounded subset. Then, $\mathbf{P}_T^F = (\theta_T \circ \Pi_T^F) \in CJ(\overline{U}, \mathbf{R}^n)$. Moreover, if $x \notin \mathbf{P}_T^F(x)$ for all $x \in \partial U$ and $deg(i - \mathbf{P}_T^F, \overline{U}) \neq 0$, then periodic problem (2.1), (2.5) has a solution.*

Notice that the properties of the translation multioperator and its applications to the periodic problem for differential inclusions of various types are described, e.g., in monographs [64, 80].

Our next target is to reduce periodic problem (2.1), (2.5) to a fixed point problem for an integral multioperator in an appropriate functional space. This method also allows to use the topological tools for solving the periodic problem and it can be considered as the base for the construction of the guiding function method.

In the sequel we assume that F is an L^1-upper Carathéodory multimap with α-sublinear growth. Let us recall (see Sect. 1.1.2) that, in this situation the superposition multioperator $\mathscr{P}_F : C([0, T]; \mathbf{R}^n) \to P(L^1([0, T]; \mathbf{R}^n))$ is well defined.

Let us consider the integral operator $j : L^1([0, T]; \mathbf{R}^n) \longrightarrow C([0, T]; \mathbf{R}^n)$,

$$j(f)(t) = \int_0^t f(s) \, ds.$$

It is an easy exercise to verify, by applying Proposition 1.5 and the classical Ascoli–Arzelá theorem, that the composition $j \circ \mathscr{P}_F : C([0, T]; \mathbf{R}^n) \to P(C([0, T]; \mathbf{R}^n))$ is closed and transfers each bounded subset $\Omega \subset C([0, T]; \mathbf{R}^n)$ onto a relatively compact set $j \circ \mathscr{P}_F(\Omega)$.

As the consequence of the above result and Proposition 1.5, we have the following assertion

Corollary 2.1. *The composition* $j \circ \mathscr{P}_F : C([0, T]; \mathbf{R}^n) \longrightarrow Kv(C([0, T]; \mathbf{R}^n))$ *is completely u.s.c. multimap.*

Denote $\mathscr{C} = C([0, T]; \mathbf{R}^n)$. The simplest integral multioperator that can be used to search for T-periodic solutions seems to be the following one:

$$J_T : \mathscr{C} \to Kv(\mathscr{C}),$$
$$J_T(x) = x(T) + j \circ \mathscr{P}_F(x).$$

The next assertion can be easily verified.

Theorem 2.2. *Fixed points of the multioperator* J_T *coincide with solutions of periodic problem (2.1), (2.5).*

From the properties of the composition $j \circ \mathscr{P}_F(x)$, mentioned above, it clearly follows that the multioperator J_T is completely u.s.c. .

So, the topological degree theory can be applied to this multioperator and we can formulate the following general principle.

Theorem 2.3. *Let* $U \subset \mathscr{C}$ *be a bounded open set. If* $deg(i - J_T, \overline{U}) \neq 0$, *then periodic problem (2.1), (2.5) has a solution in* U.

This result is one of the cornerstones on which *the method of guiding functions* can be built in its classical version. Let us outline briefly its main features.

First of all, let us describe the a priori boundedness property of solutions to Cauchy problem (2.1), (2.2). We need the following slightly modified assertion on integral inequalities known as the Gronwall lemma (see, e.g., [72], Sect. III.1.1)

Lemma 2.1 (Gronwall Lemma). *Let* $u, v : [a, b] \to \mathbf{R}$ *be nonnegative functions, u be summable, and v be continuous; $C \geq 0$ be a constant such that*

$$v(t) \leq C + \int_a^t u(s) v(s) \, ds, \quad a \leq t \leq b.$$

Then

$$v(t) \leq C e^{\int_a^t u(s) ds}, \quad a \leq t \leq b.$$

Lemma 2.2. *The set of solutions to Cauchy problem (2.1), (2.2) is a priori bounded.*

Proof. Each solution to problem (2.1), (2.2) has the form

$$x(t) = x_0 + \int_0^t f(s)\, ds,$$

where $f \in \mathscr{P}_F(x)$. Then we have the following estimate for the continuous function $v(t) = |x(t)|$, where $|\cdot|$ denotes the norm in $\mathbf{R}^n$:

$$v(t) \le |x_0| + \int_0^t |f(s)|\, ds \le |x_0| + \int_0^t \alpha(s)(1 + |x(s)|)\, ds \le$$

$$\le |x_0| + \int_0^T \alpha(s)\, ds + \int_0^t \alpha(s)\, v(s)\, ds.$$

Applying Lemma 2.1 we obtain

$$|x(t)| \le C e^{\int_0^T \alpha(s)ds}, \quad t \in [0, T],$$

where $C = |x_0| + \int_0^T \alpha(s)\, ds.$ $\square$

Let us introduce the following notion.

Definition 2.1 (cf. [90]). A point $x_0 \in R^n$ is called a *T-non-recurrence point* of trajectories of differential inclusion (2.1) if for each solution x emanating from x_0 the following condition holds:

$$x(t) \ne x_0, \quad \forall t \in (0, T]. \tag{2.6}$$

The following assertion plays a key role in the justification of the method of guiding functions. For simplicity, we restrict ourselves to the case when the right-hand side of inclusion (2.1) is u.s.c.

Theorem 2.4. *Let $U \subset \mathbf{R}^n$ be a bounded open set such that each point $x \in \partial U$ is a T-non-recurrence point of trajectories of differential inclusion (2.1). Let $F : [0, T] \times \mathbf{R}^n \to Kv(\mathbf{R}^n)$ be a u.s.c. multimap with α-sublinear growth. If the multifield $R_0 : \overline{U} \to Kv(\mathbf{R}^n)$,*

$$R_0(x) = -F(0, x),$$

does not have singular points on ∂U then

$$deg(\Phi, \overline{\Omega}) = deg(R_0, \overline{U}),$$

where $\Phi = i - J_T$ is a multifield generated by the integral multioperator J_T, and Ω is a certain bounded open set in the space $\mathscr{C}$.

Proof. From Lemma 2.2 it follows that the set of all solutions of inclusion (2.1) emanating from $\overline{U}$ is bounded. Let $m > 0$ be a number such that the norm of each solution from this set is less than m. Define an open set Ω in the space $\mathscr{C}$ by the following conditions:

$$\Omega = \{x \in \mathscr{C} \mid x(0) \in U, \ \|x\| < m\}.$$

Consider the family of multimaps

$$F_\lambda(t, x) = F(\lambda t, x), \quad \lambda \in [0, 1]$$

and the family of multifields $\Psi : \bar{\Omega} \times [0, 1] \to Kv(\mathscr{C})$ defined in the following way:

$$\Psi(x, \lambda) = \Big\{z \mid z(t) = x(t) - x(T) - \lambda \int_0^t f(s)\, ds - (1 - \lambda)$$

$$\int_0^T f(s)\, ds \colon f \in \mathscr{P}_{F_\lambda}(x)\Big\}$$

It is easy to verify that the family of multifields Ψ is completely u.s.c. Let us show that this family is non-singular on $\partial\Omega \times [0, 1]$.

Suppose the contrary, i.e., let there exist a function $x_0 \in \partial\Omega$ and a number $\lambda_0 \in [0, 1]$ such that $0 \in \Psi(x_0, \lambda_0)$. It means that there exists a summable selection $f(s) \in F(\lambda_0 s, x_0(s))$ which satisfies the following equality:

$$x_0(t) = x_0(T) + \lambda_0 \int_0^t f(s)\, ds + (1 - \lambda_0) \int_0^T f(s)\, ds \qquad (2.7)$$

for each $t \in [0, T]$.

For $t = 0$ we have

$$x_0(0) = x_0(T) + (1 - \lambda_0) \int_0^T f(s)\, ds,$$

while for $t = T$ we obtain

$$\int_0^T f(s)\, ds = 0. \qquad (2.8)$$

Whence, $x_0(0) = x_0(T)$.

By taking the derivative in t in both sides of equality (2.7) we get

$$x_0'(t) = \lambda_0 f(t) \in \lambda_0 F(\lambda_0 t, x_0(t))$$

for a.e. $t \in [0, T]$.

So, x_0 is a solution of the differential inclusion

$$x'(t) \in \lambda_0 F(\lambda_0 t, x(t)).$$

Notice, that by construction of the set Ω, its boundary consists of functions of the following two types:

1. $x(0) \in \partial U$;
2. $x(0) \in U$, $\|x\| = m$.

Consider two cases:

(a) $\lambda_0 = 0$,
(b) $\lambda_0 \neq 0$.

Case (a). Let $\lambda_0 = 0$, then $x_0(t) \equiv x_0$ for each $t \in [0, T]$, $f(t) \in F(0, x_0)$ for a.e. $t \in [0, T]$ and from (2.8) we obtain $0 \in F(0, x_0)$.

The function x_0, being a constant, can not be a function of the first type since, by condition, the multifield R_0 has no singular points on ∂U.

On the other hand, the function x_0 can not be a second type function, since $\|x_0\| < m$ by construction of the set Ω.

Case (b). Now, let $\lambda_0 \neq 0$. Consider the function $z_0(t) = x_0\left(\frac{t}{\lambda_0}\right)$. Then, for a.e. $t \in [0, \lambda_0 T]$ we have

$$z_0'(t) = \frac{1}{\lambda_0}x_0'\left(\frac{t}{\lambda_0}\right) = f\left(\frac{t}{\lambda_0}\right) \in F\left(t, x_0\left(\frac{t}{\lambda_0}\right)\right) = F(t, z_0(t)).$$

So, the function z_0 is a solution of differential inclusion (2.1) on the interval $[0, \lambda_0 T]$. According to the global existence theorem, we can extend it on the whole interval $[0, T]$.

Notice that the function x_0 can not be the first type function. Indeed, from

$$x_0(0) = z_0(0) = x_0(T) = z_0(\lambda_0 T),$$

it follows that inclusion (2.1) has a solution z_0 such that $z_0(0) \in \partial U$ and $z_0(0) = z_0(\lambda_0 T)$, contrary to the assumption that trajectories emanating from ∂U are T-non-recurrenting.

On the other hand, x_0 can not be the second type function. It follows from

$$|x_0| \leq |z_0| < m,$$

since z_0 is a solution of inclusion (2.1) emanating from the set U.

So, the family of multifields Ψ realizes the homotopy of multifields

$$\Psi_1 = \Phi = i - J_T,$$

and

$$\Psi_0(x) = i - \Gamma_0(x),$$

where the multioperator $\Gamma_0 : \overline{\Omega} \to Kv(\mathscr{C})$ is defined by the relation

$$\Gamma_0(x) = x(T) + \int_0^T F(0, x(s))\, ds.$$

This multioperator acts into the finite dimensional subspace $C_{[0,T]}^n$ of constant functions being naturally isomorphic to $\mathbf{R}^n$. By using the restriction property of the topological degree we obtain

$$deg(\Psi_0; \overline{\Omega}) = deg(\Psi_0 \mid_{\mathbf{R}^n}, \overline{U}).$$

It is easy to see that the multifield $\hat{\Psi}_0 = \Psi_0 \mid_{\mathbf{R}^n}$ is defined by the relations

$$\hat{\Psi}_0(x) = -\int_0^T F(0, x)\, ds = -T \cdot F(0, x).$$

Then we finally obtain

$$deg(\Phi, \overline{\Omega}) = deg(\Psi_0, \overline{\Omega}) = deg(-F(0, \cdot), \overline{U}) = deg(R_0, \overline{U}). \qquad \square$$

The following assertion on the existence of a periodic solution immediately follows from the proved result.

Corollary 2.2. *In conditions of Theorem 2.4, let*

$$deg(R_0, \overline{U}) \neq 0.$$

Then, differential inclusion (2.1) has a T-periodic solution.

Now we apply Theorem 2.4 to justify the classical version of the method of guiding functions. Let us introduce the necessary notions.

Definition 2.2. A continuously differentiable function $v : \mathbf{R}^n \to \mathbf{R}$ is called *non-degenerate potential* if its gradient is non-zero outside a certain ball centered at the origin, i.e., there exists $r_v > 0$ such that

$$grad\, v\,(x) = \left\{ \frac{\partial v\,(x)}{\partial x_1}, \frac{\partial v\,(x)}{\partial x_2}, \ldots, \frac{\partial v\,(x)}{\partial x_n} \right\} \neq 0,$$

for each $x \in \mathbf{R}^n$, $|x| \geq r_v$.

From the properties of the topological degree (see, e.g., [25, 38, 80, 89, 95]) it follows that the degree of the gradient of a non-degenerate potential

$$deg\,(grad\, v\,(x)\,, B_r)$$

on the closed ball $B_r \subset \mathbf{R}^n$ of radius $r \geq r_v$, centered at the origin, does not depend on r. This generic value of the degree is called *the index of a non-degenerate potential* and it is denoted as *ind* v.

As an example of potential with non-zero index we can consider a non-degenerate potential v satisfying the coercivity condition

$$\lim_{|x| \to \infty} |v\,(x)| \to \infty. \tag{2.9}$$

(see [95]).

Other examples of potential with non-zero index can be found in [90, 95].

Definition 2.3. A non-degenerate potential v is called a strict guiding function for differential inclusion (2.1) if

$$\langle grad\, v\,(x)\,, y \rangle > 0 \tag{2.10}$$

for all $y \in F\,(t, x)$, $0 \leq t \leq T$, $|x| \geq r_v$.

From this definition it follows immediately that if v is a strict guiding function of inclusion (2.1) then the field $-grad\, v$ and the multifield R_0 does not allow opposite directions on spheres S_r of the radius $r \geq r_v$, and hence, by Lemma 1.5

$$deg\,(R_0, B_r) = (-1)^n\, ind\, v. \tag{2.11}$$

(We have used the known property of the degree of single-valued fields: $deg\,(-\varphi, S) = (-1)^n\, deg\,(\varphi, S)$, see, e.g., [95]).

We can formulate now the following condition for the existence of a periodic solution.

Theorem 2.5. *Let* $F : [0, T] \times \mathbf{R}^n \to Kv\,(\mathbf{R}^n)$ *be an u.s.c. multimap with* α-*sublinear growth. If, for differential inclusion (2.1), there exists a strict guiding function* v *of non-zero index, then the inclusion has a* T-*periodic solution.*

To prove this assertion, we need the following technical result

Lemma 2.3. *Let*

$$r_0 = (r_v + \int_0^T \alpha(s)\, ds) e^{\int_0^T \alpha(s)\, ds} \qquad (2.12)$$

where α is the function from the sublinear growth condition (F3').

If x is a solution of inclusion (2.1) with initial condition $|x(0)| > r_0$, then $|x(t)| > r_v$ for all $t \in [0, T]$.

Proof. Indeed, let there exists $t_0 \in [0, T]$ such that $|x(t_0)| \leq r_v$. For $t \in [0, t_0]$, define

$$y(t) = x(t_0 - t), \quad \beta(t) = \alpha(t_0 - t), \quad G(t, x) = -F(t_0 - t, x).$$

It is clear that

$$y'(t) \in G(t, y(t)).$$

Since $\|G(t, x)\| \leq \beta(t)(1 + |x|)$ for a.e. $t \in [0, t_0]$, applying Lemma 2.2, we obtain

$$|y(t)| \leq \left(|y(0)| + \int_0^{t_0} \beta(s)\, ds \right) e^{\int_0^{t_0} \beta(s)\, ds} \leq r_0$$

for all $t \in [0, t_0]$. So, $|x(0)| = |y(t_0)| \leq r_0$ and we get the contradiction. $\square$

Proof (of Theorem 2.5). Notice that, for each $r > r_0$, the sphere S_r consists of T-non-recurrence points of inclusion (2.1). Indeed, if x is a solution of (2.1) such that $x(0) \in S_r$, then from Lemma 2.3 it follows that $|x(t)| > r_v$ for all $t \in [0, T]$. Then for each $t \in (0, T]$ we have

$$v(x(t)) - v(x(0)) = \int_0^t \langle grad\, v(x(s)), x'(s)\rangle ds > 0, \qquad (2.13)$$

and therefore relation (2.6) follows.

To conclude the proof, it remains to apply relation (2.11) and Corollary 2.2. $\square$

This version of the method of guiding functions for differential inclusions allows extensions in various directions. Let us discuss some of them.

First of all, we extend the notion of guiding function as well as the class of differential inclusions to which the MGF can be applied.

Definition 2.4. A non-degenerate potential v is called a guiding function for the differential inclusion (2.1) if

$$\langle grad\, v(x), y\rangle \geq 0$$

for all $y \in F(t, x)$, $0 \leq t \leq T$, $|x| \geq r_v$.

Proposition 2.3. Let $F : [0, T] \times \mathbf{R}^n \to Kv(\mathbf{R}^n)$ be an L^1-upper Carathéodory multimap with α-sublinear growth. If differential inclusion (2.1) admits a guiding function v of non-zero index, then the inclusion (2.1) has a T-periodic solution.

Proof. STEP 1   Let us show that the assertion is true for an u.s.c. F.
For $k = 0, 1, 2, \ldots$, set

$$M_k = \sup\{|grad\, v(x)| : x \in B(k)\},$$

where $B(k) \subset \mathbf{R}^n$ denotes a closed ball of radius k, centered at the origin.
Define the function $\eta : \mathbf{R}^n \to \mathbf{R}$ as

$$\eta(x) = 1 + (|x| - k)M_{k+2} + (k + 1 - |x|)M_{k+1}, \quad k \le |x| \le k+1.$$

It is easy to see that the function η is continuous and satisfies the condition

$$\eta(x) \ge \max\{1, |grad\, v(x)|\} \quad \text{for all} \quad x \in \mathbf{R}^n.$$

So, the map $g : \mathbf{R}^n \to \mathbf{R}$,

$$g(x) = \frac{grad\, v(x)}{\eta(x)}$$

is continuous and satisfies the condition $|g(x)| \le 1$ for all $x \in \mathbf{R}^n$.
For any sequence $\{\epsilon_m\}$ of positive numbers, consider the corresponding sequence
of auxiliary differential inclusions

$$x'(t) \in F(t, x(t)) + \epsilon_m g(x(t)). \qquad (2.14)$$

It is clear that the right-hand side of each inclusion (2.14) is u.s.c. with
α-sublinear growth.
For each $|x| \ge r_v$ and $y \in F(t, x)$ we have

$$\langle grad\, v(x), y + \epsilon_m g(x)\rangle = \langle grad\, v(x), y\rangle + \epsilon_m \frac{\langle grad\, v(x), grad\, v(x)\rangle}{\eta(x)} > 0$$

and, by Theorem 2.5 inclusion (2.14) has a T-periodic solution x_m for each
$\epsilon_m > 0$. Tending the sequence $\{\epsilon_m\}$ to zero, we obtain the desired solution of (2.1)
as a limit point of the sequence $\{x_m\}$.
STEP 2   Now, assume that F is an L^1-upper Carathéodory multifunction with
α-sublinear growth.
From [39], Sect. 5, it follows that we can assume, w.l.o.g., that the multimap F
is bounded and, then we have the following result (see [39], Proposition 5.1)

Lemma 2.4. *For each $\epsilon > 0$ there exists a multimap $F_\epsilon : [0, T] \times \mathbf{R}^n \to Kv(\mathbf{R}^n)$
such that:*

1) $F_\epsilon(t, x) \subset F(t, x)$, $(t, x) \in [0, T] \times \mathbf{R}^n$;
*2) there exists a closed subset $J_\epsilon \subset [0, T]$ with $\mu([0, T]\backslash J_\epsilon) \le \epsilon$ such that $F_{|J_\epsilon \times \mathbf{R}^n}$
is u.s.c.*

For a sequence $\{\epsilon_m\}$ of positive numbers tending to zero, let us take the corresponding sequence of multimaps $F_{\epsilon_m} : [0, T] \times \mathbf{R}^n \to Kv(\mathbf{R}^n)$ satisfying, for each ϵ_m conditions (1), (2) of the above lemma.

For each ϵ_m, let $P_m : [0, T] \multimap J_{\epsilon_m}$ be a metric projection and the multimap $\tilde{F}_{\epsilon_m} : [0, T] \to Kv(\mathbf{R}^n)$ be defined as

$$\tilde{F}_{\epsilon_m}(t, x) = \overline{co}\, F_{\epsilon_m}(P_m(t), x) \ .$$

Since each metric projection P_m is closed and, hence, u.s.c., from Propositions 1.7, 1.8 and 1.12 it follows that each multimap $\tilde{F}_{\epsilon_m}$ is u.s.c.

Furthermore, it easy to see that for each $\tilde{F}_{\epsilon_m}$ relation in Definition 2.4 is fulfilled and, according to Step 1, each differential inclusion

$$x'(t) \in \tilde{F}_{\epsilon_m}(t, x(t))$$

has a T-periodic solution x_m.

Tending m to infinity we obtain a T-periodic solution x as a limit point of the sequence $\{x_m\}$. $\qquad\square$

To obtain further generalizations, let us introduce the following notion

Definition 2.5. A non-degenerate potential v is called a weak guiding function for differential inclusion (2.1) if

$$\langle grad\, v(x), y \rangle \geq 0 \text{ for at least one } y \in F(t, x) \text{ and for all } t \in [0, T], |x| \geq r_v$$
$$(2.15)$$

Now we can formulate the most general result concerning the application of the MGF to periodic problem (2.1), (2.5).

Theorem 2.6. *Let* $F : [0, T] \times \mathbf{R}^n \to Kv(\mathbf{R}^n)$ *be an* L^1-*upper Carathéodory multimap with* α-*sublinear growth. If differential inclusion (2.1) admits a weak guiding function of non-zero index, then the inclusion has a* T-*periodic solution.*

Proof. Let us define the multimap $B : \mathbf{R}^n \to Cv(\mathbf{R}^n)$ as

$$B(x) = \{y \in \mathbf{R}^n : \langle \gamma(x) grad\, v(x), y \rangle \geq 0\}$$

where

$$\gamma(x) = \begin{cases} 0, & |x| \leq r_v, \\ 1, & |x| > r_v \end{cases}$$

It is easy to verify that the multimap B is closed. So, by applying Lemma 1.9 we can see that the multimap $F^B : [0, T] \times \mathbf{R}^n \to Kv(\mathbf{R}^n)$,

$$F^B(t, x) = F(t, x) \cap B(x)$$

is well defined and is L^1-upper Carathéodory with α-sublinear growth. Moreover, for the multimap F^B condition (2.15) is fulfilled for all $|x| \geq r' > r_v$ and $y \in F^B(t, x)$, where r' is an arbitrary number. Therefore v is a guiding function for the differential inclusion

$$x'(t) \in F^B(t, x)).\tag{2.16}$$

From Proposition 2.3 it follows that inclusion (2.16) has a solution, implying the result.   □

Corollary 2.3. *let* $F : [0, T] \times \mathbf{R}^n \to Kv(\mathbf{R}^n)$ *be an* L^1*-upper Carathéodory multimap with* α*-sublinear growth. If the differential inclusion (2.1) admits a weak guiding function* v *satisfying the coercivity condition (2.9), then the inclusion has a* T*-periodic solution.*

2.2 Non-smooth Guiding Functions

In the previous section, following the classical works on the method of guiding functions, we supposed the guiding function smooth on the whole space. This condition may be onerous, for example, in situations where the guiding potentials are different on different domains. In this case it is natural to take as the potential, defined on the whole space, the maximum of all potentials but, this new function may be non-smooth, in general. In this section we describe the extension of the notion of a guiding function to the non-smooth case. Notice, in this connection, that non-differentiable Liapunov functions are effectively used in the stability theory (see, e.g., [133]).

To deal with such potentials, let us recall some notions of non-smooth analysis (see, e.g., [34]).

Let X be a real Banach space endowed with the norm $\|\cdot\|$. The dual space is denoted by X^* and the notation $\langle\cdot, \cdot\rangle$ means the duality pairing between X^* and X.

Definition 2.6. A function $V : X \to \mathbf{R}$ is called locally Lipschitz if for each point $x \in X$ there exist a neighborhood U of x and a constant $C > 0$ such that

$$|V(y) - V(z)| \leq C \, \|y - z\|, \quad \forall y, z \in U.$$

Remark 2.1. A convex and continuous function $V : X \to \mathbf{R}$ is locally Lipschitz. More generally, a convex function $V : X \to \mathbf{R}$ which is bounded above on a neighborhood of some point is locally Lipschitz (see [34]).

Definition 2.7. Let $V : X \to \mathbf{R}$ be a locally Lipschitz function. For $x \in X$ and $v \in X$, the generalized derivative $V^0(x; v)$ of V at a point x in the direction v is given by the formula

$$V^0(x; v) = \varlimsup_{\substack{z \to x \\ t \to 0+}} \frac{V(z + tv) - V(z)}{t}.\tag{2.17}$$

Definition 2.8. A locally Lipschitz function $V : X \to \mathbf{R}$ is called regular at a point $x \in X$ if the usual directional derivative

$$V'(x; v) = \lim_{t \to 0+} \frac{V(x + tv) - V(x)}{t}$$

exists for each $v \in X$ and is equal to $V^0(x; v)$.

Remark 2.2. A convex and continuous function $V : X \to \mathbf{R}$ is regular.

Now we can introduce the following notion.

Definition 2.9. The Clarke's generalized gradient of a locally Lipschitz function $V : X \to \mathbf{R}$ at a point $x \in X$ is defined as the set $\partial V(x) \subset X^*$ in the following way:

$$\partial V(x) = \{z \in X^* : \langle z, v \rangle \le V^0(x; v) \quad \text{for all} \ v \in X\}. \tag{2.18}$$

Notice that from the classic Hahn–Banach theorem it follows that $\partial V(x) \neq \emptyset$.

Remark 2.3. If a function $V : X \to \mathbf{R}$ is continuously differentiable, then $\partial V(x) = V'(x)$ for all $x \in X$, where $V'(x)$ denotes the Fréchet derivative of V at x.

Remark 2.4. If $X = \mathbf{R}^n$ and a function $V : \mathbf{R}^n \to \mathbf{R}$ is convex and continuous, then the Clarke's generalized gradient $\partial V(x)$ coincides with the subdifferential of V at x in the sense of convex analysis, i.e.

$$\partial V(x) = \{z \in \mathbf{R}^n \ : < z, y - x > \le V(y) - V(x), \quad \forall y \in \mathbf{R}^n\}.$$

Lemma 2.5 (see [34]). *If $X = \mathbf{R}^n$, then, for a given locally Lipschitz function $V : \mathbf{R}^n \to \mathbf{R}$, the multimap $\partial V : \mathbf{R}^n \to P(\mathbf{R}^n)$ has compact convex values and is u.s.c.*

In the sequel, we will use the following result (see [40])

Lemma 2.6. *Let a function $V : \mathbf{R}^n \to \mathbf{R}$ be regular, $x : [a, b] \to \mathbf{R}^n$ an absolutely continuous function. Then, the function $t \to V(x(t))$, $t \in [a, b]$ is absolutely continuous and*

$$V(x(t'')) - V(x(t')) = \int_{t'}^{t''} V^0(x(s), x'(s)) \, ds \quad \text{for each} \ \ t', t'' \in [a, b]$$

We start with the following notions

Definition 2.10. A regular function $V : \mathbf{R}^n \to \mathbf{R}$ is called a non-degenerate non-smooth potential if there exists $r_V > 0$ such that

$$0 \notin \partial V(x) \quad \text{for all} \quad \|x\| \ge r_V.$$

Definition 2.11. A regular function $V : \mathbf{R}^n \to \mathbf{R}$ is called a direct potential if there exists $r_V > 0$ such that

$$< v, v' > > 0 \quad \text{for all} \ v, v' \in \partial V(x), \ \|x\| \ge r_V.$$

It is obvious that each direct potential is non-degenerate, but the converse is not true in general.

Analogously to the classical case, for a non-degenerate non-smooth potential V, the topological degree of the multifield ∂V, $deg(\partial V, B_r)$ on each ball centered at the origin $B_r = B(0, r) \subset \mathbf{R}^n$ with $r \geq r_V$ is constant and is called *the index* of V and denoted by *ind V*.

The following statement gives an example of a non-smooth potential with non-zero index (see [21]).

Proposition 2.4. *If $V : \mathbf{R}^n \to \mathbf{R}$ is a direct potential, satisfying the coercivity condition* $\lim_{|x| \to \infty} V(x) = +\infty$, *then ind $V = 1$.*

Now, our target is to extend the method of guiding functions to the case of non-smooth potentials.

Definition 2.12. A non-degenerate non-smooth potential $V : \mathbf{R}^n \to \mathbf{R}$ is called a strict non-smooth guiding function for differential inclusion (2.1) if

$$< v, y >> 0 \quad \text{for all} \quad v \in \partial V(x), \ y \in F(t, x), \quad t \in [0, T], \tag{2.19}$$

where $\|x\| \geq r_V$.

Definition 2.13. A direct potential $V : \mathbf{R}^n \to \mathbf{R}$ is called a non-smooth guiding function for differential inclusion (2.1) if

$$< v, y > \geq 0 \quad \text{for all} \ y \in F(t, x), \quad v \in \partial V(x), t \in [0, T], |x| \geq r_V \tag{2.20}$$

The main result in this section is the following assertion.

Theorem 2.7. *If the right-hand part of differential inclusion (2.1) is L^1-upper Carathéodory with α-sublinear growth and the inclusion admits a non-smooth guiding function of a non-zero index, then periodic problem (2.1), (2.5) has a solution.*

Proof. STEP 1 Let us suppose F is u.s.c. and the guiding function V is strict. Observing that condition (2.19) implies that the multifields $R_0(x) = -F(0, x)$ and $\partial V(x)$ does not allow opposite directions on spheres S_r, $r \geq r_V$, we conclude, by Lemma 1.5 that

$$deg(R_0, B_r) = (-1)^n indV \neq 0.$$

To apply Corollary 2.2, it is sufficient to verify that each sphere $S_r, r > r_0$, where r_0 is defined by (2.12), consists of T-non-recurrence points of inclusion (2.1). Indeed, let x be a solution of inclusion (2.1) such that $x(0) \in S_r$ and hence $|x(t)| > r_V$ for all $t \in [0, T]$ (see Lemma 1.3). Notice that the multifunction $t \multimap \partial V(x(t)), t \in [0, T]$ is u.s.c. and hence measurable. So, by Proposition 1.14(c) it has a measurable selection $z(t) \in \partial V(x(t))$ for a.e. $t \in [0, T]$.

Then, applying Lemma 2.6 we have, for each $t \in [0, T]$.

$$V(x(t)) - V(x(0)) = \int_0^t V^0(x(s), x'(s))\, ds \geq \int_0^t \langle z(s), x'(s) \rangle\, ds > 0$$

implying the result.

STEP 2   Now, let F be u.s.c. and V a non-smooth guiding function, i.e. it is direct and satisfies condition (2.20).

Taking $|\partial V(x)| = \sup\{|v| : v \in \partial V(x)\}$ and defining $M_k = \sup\{|\partial V(x)| : x \in B_k\}$, $k = 0, 1, \ldots$ we obtain, similarly to STEP 1 in the proof of Proposition 2.3, a continuous function $\eta : \mathbf{R}^n \to \mathbf{R}$ satisfying $\eta(x) \geq \max\{1, |\partial V(x)|\}$. So, the multifunction $G : \mathbf{R}^n \to Kv(\mathbf{R}^n)$ defined by

$$G(x) = \frac{1}{\eta(x)} \partial V(x)$$

is u.s.c. and satisfies $|G(x)| \leq 1$, $\forall x \in \mathbf{R}^n$.

For a sequence of positive numbers $\{\epsilon_m\}$ tending to zero, let us define the corresponding sequence of differential inclusions

$$x'(t) \in F_m(t, x(t)) := F(t, x(t)) + \epsilon_m G(x(t)) \tag{2.21}$$

It is clear that each F_m is u.s.c. with the α-sublinear growth.

Further, each $z \in F_m(t, x)$ has the form $z = y + \frac{\epsilon_m}{\eta(x)} v'$, where $v' \in \partial V(x)$. So, for each $v \in \partial V(x)$ we have

$$\langle v, z \rangle = \langle v, y \rangle + \frac{\epsilon_n}{\eta(x)} \langle v, v' \rangle > 0$$

since the potential V is direct.

By Step 1, for each m, differential inclusion (2.21) has a solution x_m and we obtain the solution x as a limit point of the sequence $\{x_m\}$.

STEP 3   The approach to the case when F is L^1-upper Carathéodory can be made as in Step 2 of the proof of Proposition 2.3.   □

Corollary 2.4. *If inclusion (2.1) admits a non-smooth coercive guiding function, then periodic problem (2.1), (2.5) has a solution.*

2.3 Integral Guiding Functions

It is easy to see that the direct application of the MGF in its classical interpretation to the periodic problem for functional differential equations and inclusions meets difficulties. To avoid them, in this section we consider the notion of integral guiding function, first introduced by A. Fonda (see [55]) and then developed in the works [84–86].

Let us start with the necessary preliminaries.

Given $\tau > 0$, let us denote the space of continuous functions $C([-\tau, 0]; \mathbf{R}^n)$ by the symbol $\mathscr{C}$. For a function $x(\cdot) \in C([-\tau, T]; \mathbf{R}^n)$, $T > 0$ the symbol $x_t \in \mathscr{C}$, $t \in [0, T]$ denotes the function defined as $x_t(\theta) = x(t + \theta)$, $\theta \in [-\tau, 0]$.

We consider the periodic problem for a functional differential inclusion of the following form:

$$x'(t) \in F(t, x_t) \quad \text{a.e. } t \in [0, T], \tag{2.22}$$

$$x(0) = x(T), \tag{2.23}$$

assuming that $F : \mathbf{R} \times \mathscr{C} \to Kv(\mathbf{R}^n)$ is a T-periodic L^1-upper Carathéodory multimap satisfying the α-sublinear growth condition.

As earlier, by a solution of problem (2.22), (2.23) we mean an absolutely continuous function $x(\cdot)$, satisfying the periodicity condition (2.23) and the inclusion (2.22) a.e. on $[0, T]$.

Denote by C_T the space of continuous T-periodic functions $x : \mathbf{R} \to \mathbf{R}^n$ with the norm $\|x\|_C = \sup_{t \in [0,T]} |x(t)|$ and by L_T^1 the space of summable T-periodic functions $f : \mathbf{R} \to \mathbf{R}^n$ with the norm $\|f\|_{L^1} = \frac{1}{T} \int_0^T |f(t)| \, dt$.

For each $x \in C_T$ we will also consider the norm $\|x\|_2 = \left(\int_0^T |x(t)|^2 dt \right)^{1/2}$.

Now we can introduce the following notion.

Definition 2.14. A regular function $V : \mathbf{R}^n \to \mathbf{R}$ is said to be a non-smooth integral guiding function for problem (2.22)–(2.23) if there exists $N > 0$ such that for each absolutely continuous function $x \in C_T$ with $\|x\|_2 \geq N$ and $|x'(t)| \leq \|F(t, x_t)\|$ for a.e. $t \in [0, T]$ we have

$$\int_0^T \langle v(s), f(s) \rangle \, ds > 0$$

for all summable selections $v(s) \in \partial V(x(s))$ and $f(s) \in F(s, x_s)$.

Remark 2.5. As we know, the generalized gradient ∂V is a u.s.c. multimap and so, the multifunction $s \to \partial V(x(s))$ is u.s.c. and, hence, bounded and measurable. So the multifunction $\partial V(x(s))$ admits a measurable selection.

The following assertion holds.

Theorem 2.8. *Let $V : \mathbf{R}^n \to \mathbf{R}$ be a non-smooth integral guiding function of problem (2.22)–(2.23). If V is a non-degenerate potential and ind $V \neq 0$, then problem (2.22)–(2.23) has a solution.*

Proof. We use the coincidence degree theory (see Sect. 1.3). Consider the following operators:

$$L : \text{dom } L := \{x \in C_T : x \text{ is absolutely continuous}\} \subset C_T \to L_T^1,$$

$$Lx = x',$$

the superposition multioperator $G = \mathscr{P}_F : C_T \to P(L_T^1)$ and the projection $\Pi : L_T^1 \to R^n$, given as

$$\Pi f = \frac{1}{T} \int\limits_0^T f(s)\, ds.$$

It is easy to verify (see Proposition 1.17) that multioperators ΠG and $K_{P,Q} G$ are compact and upper semicontinuous.

Let us mention that periodic problem (2.22)–(2.23) is reduced to the existence of a coincidence point $x \in dom\, L$ for the pair (L, G) :

$$Lx \in G(x).$$

Moreover, for $\lambda \in (0, 1]$, an arbitrary solution $x \in dom\, L$ of the inclusion

$$Lx \in \lambda G(x)$$

satisfies the relations

$$x'(t) \in \lambda F(t, x_t),$$
$$x(0) = x(T).$$

It means that $x(\cdot)$ is an absolutely continuous function such that $x'(t) = \lambda f(t)$ for a.e. $t \in [0, T]$, where $f \in \mathscr{P}_F(x)$.

Then, applying Lemma 2.6 we obtain, for each summable selection $v(s) \in \partial V(x(s))$

$$\int_0^T \langle v(s), f(s) \rangle\, ds = \frac{1}{\lambda} \int_0^T \langle v(s), x'(s) \rangle\, ds$$

$$\leq \frac{1}{\lambda} \int_0^T V^0(x(s), x'(s))\, ds = \frac{1}{\lambda}\, (V(x(T)) - V(x(0))) = 0\,,$$

and hence

$$\|x\|_2 < N.$$

Then, from α-sublinear growth condition of F it follows that there exists a constant $M' > 0$ such that $\|x'\|_2 < M'$. So, there exists also $M > 0$ such that

$$\|x\|_C < M$$

Denote by U the ball $B_r \subset C_T$ centered at the origin with the radius $r = \max\{r_V, M, NT^{-1/2}\}$. Then we have

$$Lx \notin \lambda G(x)$$

for all $x \in \partial U, \lambda \in (0, 1]$.

Further, take an arbitrary point $u \in \partial U \cap Ker\, L$. Since u is a constant function satisfying $\|u\|_C \geq NT^{-1/2}$, from the definition of the integral guiding function we obtain

$$\int_0^T \langle v(s), f(s) \rangle\, ds > 0$$

for all measurable selections $v(s) \in \partial V(u)$, $f(s) \in F(s, u)$. Taking a constant selection $v(s) \equiv v \in \partial V(u)$ we have

$$\int_0^T \langle v, f(s) \rangle\, ds = \langle v, \int_0^T f(s)\, ds \rangle = T \langle v, \Pi f \rangle > 0$$

for all $v \in \partial V(u)$, and hence

$$\langle v, y \rangle > 0$$

for all $v \in \partial V(u)$, $y \in \Pi G(u)$.

This means that the multifields $\partial V(u)$ and $\Pi G(u)$ are homotopic on $\partial U \cap Ker\, L$ and hence

$$\deg_{Ker\, L}(\Pi G|_{\overline{U}_{\mathrm{ker}\, l}}, \overline{U}_{Ker\, L}) = \deg(\partial V, \overline{U}_{Ker\, L}) \neq 0.$$

So, all conditions of Theorem 1.2 are fulfilled and so the operators L and G have a coincidence point in the ball U and hence problem (2.22)–(2.23) has a solution in the same ball.   □

As an example, we consider the periodic problem for a gradient functional differential inclusion of the following form:

$$x'(t) \in \partial G(x(t)) + F(t, x_t) \tag{2.24}$$

$$x(0) = x(T), \tag{2.25}$$

where the multimap F is T-periodic in the first argument and is L^1-upper Carathéodory with α-sublinear growth and ∂G is the generalized gradient of a regular function $G : \mathbf{R}^n \to \mathbf{R}$.

Theorem 2.9. *Suppose that the following conditions are satisfied:*

(A1) there exist constants $\varepsilon > 0$, $K > 0$ and $\beta \geq 1$ such that

$$|g| \geq \varepsilon |u|^{\beta} - K$$

for all $g \in \partial G(u)$, $u \in \mathbf{R}^n$;

(A2)

$$\lim_{\|x\|_2 \to \infty} \frac{\|\mathscr{P}_F(x)\|_2}{\|x\|_2^{\beta}} < \varepsilon T^{(1-\beta)/2}$$

for absolutely continuous functions $x \in C_T$;

(*A3*) *the generalized gradient* ∂G *has a non-zero topological degree:*

$$\deg(\partial G, \overline{B}_N) \neq 0$$

   for sufficiently large $N > 0$.

Then problem (2.24)–(2.25) has a solution.

Proof. Let us demonstrate that G is a non-smooth integral guiding function for problem (2.24)–(2.25). Notice that the embedding $L^{2\beta} \subset L^2$ gives the following estimation for each absolutely continuous function $x(\cdot) \in C_T$ and every summable selection $g(t) \in \partial G(x(t))$:

$$\|g\|_2 \geq \varepsilon\|x\|_{2\beta}^\beta - K\sqrt{T} \geq \varepsilon T^{(1-\beta)/2}\|x\|_2^\beta - K\sqrt{T}.$$

Then for each summable selections $f \in \mathscr{P}_F(x)$ and $g(t) \in \partial G(x(t))$ we have

$$\int_0^T \langle g(s), g(s) + f(s)\rangle\, ds \geq \|g\|_2(\|g\|_2 - \|f\|_2)$$

$$\geq \|g\|_2(\|g\|_2 - \|\mathscr{P}_F(x)\|_2)$$

$$\geq \|g\|_2\Big(\varepsilon T^{(1-\beta)/2} - \frac{K\sqrt{T}}{\|x\|_2^\beta} - \frac{\|P_F(x)\|_2}{\|x\|_2^\beta}\Big)\|x\|_2^\beta > 0$$

for $\|x\|_2$ sufficiently large. $\qquad\qquad\square$

2.4 Generalized Periodic Problems

We consider, here, the application of the MGF to some generalization of the classical periodic problem for differential inclusions. We discuss also some applications to differential games and other examples including the anti-periodic problem.

   Starting from this section, in the sequel we use the symbol V for a smooth non-degenerate potential.

2.4.1 Preliminaries

Definition 2.15. Let X, Y be Banach spaces. By $J^c(X, Y)$ we denote the collection of all multimaps $F: X \to K(Y)$ that can be represented in the form of a composition

$$F = \Sigma_q \circ \cdots \circ \Sigma_1,$$

where $\Sigma_i \in J(X_{i-1}, X_i)$, $i = 1 \cdots q$, $X_0 = X$, $X_q = Y$, and X_i ($0 < i < q$) are normed spaces.

Following the construction of the topological degree for CJ-multimap given in the chapter "Introduction", let us mention that if $U \subset \mathbf{R}^n$ is an open bounded subset and $F: \overline{U} \to K(\mathbf{R}^n)$ is a J^c-multimap such that $0 \notin F(x)$ for all $x \in \partial U$, then the topological degree $deg(F, \overline{U})$ is well-defined and has all usual properties of the Brouwer topological degree.

For a non-degenerate potential $V : \mathbf{R}^n \to \mathbf{R}$ define the vector field $W_V : \mathbf{R}^n \to \mathbf{R}^n$,

$$W_V(x) = \begin{cases} grad\, V(x) & \text{if } |grad\, V(x)| \leq 1, \\ \frac{grad\, V(x)}{|grad\, V(x)|} & \text{if } |grad\, V(x)| > 1. \end{cases}$$

In the sequel we need the following result.

Lemma 2.7 (see [64, Lemma 72.8]). . *Let $r_V > 0$ be a constant for the non-degenerate potential V. Then for every $r > r_V + a$, $a > 0$, there is $t_r \in (0, a]$ such that:*
for each solution $x: [0, a] \to \mathbf{R}^n$ of the problem

$$\begin{cases} x'(t) = W_V(x(t)) \\ |x(0)| = r \end{cases}$$

the following relations hold:

$$\langle x(t) - x(0), grad\, V(x(0)) \rangle > 0, \quad \forall t \in (0, t_r].$$

$$x(t) - x(0) \neq 0, \quad \forall t \in (0, a]$$

2.4.2 The Setting of the Problem

Consider the following generalized periodic problem

$$\begin{cases} u'(t) \in F(t, u(t)), \text{ for a.e. } t \in [0, T], \\ u(T) \in M(u(0)), \end{cases} \tag{2.26}$$

with the assumptions that:

(A1) $F: [0, T] \times \mathbf{R}^n \to Kv(\mathbf{R}^n)$ is a L^1-upper Carathéodory multimap with α-sublinear growth;
(A2) $M: \mathbf{R}^n \to K(\mathbf{R}^n)$ is a J^c-multimap.

By a solution to problem (2.26) we mean an absolutely continuous function $u: [0, T] \to \mathbf{R}^n$ satisfying (2.26).

Definition 2.16. A non-degenerate potential $V: \mathbf{R}^n \to \mathbf{R}$ is said to be a guiding function for problem (2.26) if there exists $r_* > 0$ such that for every $(t, x) \in [0, T] \times \mathbf{R}^n$, $|x| \geq r_*$, the following relations hold:

(V1) $\langle grad\, V(x), y \rangle \geq 0$ for at least one point $y \in F(t, x)$;
(V2) $V(x) \geq V(w)$ for all $w \in M(x)$;
(V3) $\langle grad\, V(x), x - w \rangle \geq 0$ for at least one point $w \in M(x)$, if M has convex values, otherwise for all $w \in M(x)$.

2.4.3 Application to Differential Games

Let us mention that the class of problems having the form (2.26) is sufficiently wide. It is clear that in the case when M is the identity operator, i.e., $M(x) = x$, $\forall x \in \mathbf{R}^n$, problem (2.26) is the classical periodic problem. Consider some other examples of problems which can be represented in form (2.26).

Problem 2.1 (Differential game with a given goal set). Consider a differential game in which an object moves along the trajectories of the following differential inclusion

$$u'(t) \in F(t, u(t)), \tag{2.27}$$

where $F: [0, T] \times \mathbf{R}^n \to Kv(\mathbf{R}^n)$ is a given multimap.

It is supposed that for each initial position $u(0) = x$ a goal set $M(x) \subset \mathbf{R}^n$ is given. The game ends if, starting from a position x, the object can be moved to one of the goal positions $M(x)$ at the time T.

The game is called finite if there are an initial position x and a trajectory of (2.27) such that the game ends. Otherwise the game is called infinite.

It is clear that the finiteness of the game is equivalent to the existence of a solution of problem (2.26).

Problem 2.2 (Differential game of pursuit). In this game, two participating players A and B start moving at the same time from different initial positions along the trajectories of the differential inclusions

$$u'(t) \in G_0(t, u(t)), \tag{2.28}$$

(player A) and, respectively:

$$v'(t) \in G_1(t, v(t)), \tag{2.29}$$

(player B).
It is supposed that $G_1: [0, T] \times \mathbf{R}^n \to Kv(\mathbf{R}^n)$ is a L^1-upper Carathéodory multimap with α-sublinear growth.

The player A is considered as a pursuer whereas the player B is an evader. We assume that for each chosen initial position x of the pursuer A, the evader

B starts the game from the initial position $h(x)$ defined by a continuous function $h: \mathbf{R}^n \to \mathbf{R}^n$. The game ends if, at the moment T, the players A and B reach the same position, i.e., player A "catches up" player B at this time.

The game is called finite if there are an initial position x and trajectories of (2.28) and (2.29), respectively, such that the game ends.

Let us reduce the game to problem (2.26). To this aim, let us recall (see Proposition 2.1) that under the assumptions imposed on the multimap G_1, for each $y \in \mathbf{R}^n$ the Cauchy problem

$$\begin{cases} v'(t) \in G_1(t, v(t)), & \text{for a.e. } t \in [0, T], \\ v(0) = y, \end{cases}$$

has a solution. Moreover, if we denote by $\Pi_{G_1}(y)$ the set of all solutions, then the multimap

$$\Pi_{G_1} : \mathbf{R}^n \to K(\mathbf{R}^n) \tag{2.30}$$

is a J-multimap.

Now, let $x \in \mathbf{R}^n$ be an initial position of A. Then $h(x)$ is the initial position of B. For every $t \in [0, T]$ define an evaluation operator:

$$\theta_t : C([0, T]; \mathbf{R}^n) \to \mathbf{R}^n, \ \theta_t(u) = u(t), \tag{2.31}$$

and consider the following multimap:

$$M : \mathbf{R}^n \to K(\mathbf{R}^n), \ M(x) = \theta_T \circ \Pi_{G_1} \circ h(x).$$

It is easy to see that M is a J^c-multimap and the finiteness of the game is equivalent to the existence of a solution of the following problem

$$\begin{cases} u'(t) \in G_0(t, u(t)), \\ u(T) \in M(u(0)). \end{cases}$$

2.4.4 Existence Theorem, Corollaries and Example

Theorem 2.10. *Let conditions (A1)–(A2) hold. In addition, assume that there exists a guiding function V for problem (2.26) such that $\operatorname{ind} V \neq 0$. Then problem (2.26) has a solution.*

Proof. Set $r = \max\{r_V, r_*\}$, where r_V is the constant for the non-degenerate potential V and r_* is the constant from Definition 2.16.

Let M be a convex-valued multimap. Define a multimap

$$B: \mathbf{R}^n \to P(\mathbf{R}^n),$$

$$B(x) = \{y \in \mathbf{R}^n : \langle x - y, \varphi(x) \operatorname{grad} V(x) \rangle \geq 0\},$$

where $\varphi(x) = 0$ if $|x| \leq r$ and $\varphi(x) = 1$ if $|x| > r$.

It is easy to see that B is a closed multimap with convex values. Therefore, the multimap

$$M_B: \mathbf{R}^n \to K(\mathbf{R}^n), \ M_B(x) = M(x) \cap B(x),$$

is a J-multimap, and hence it is a J^c-multimap (see Proposition 1.9).

Moreover, for every $x \in \mathbf{R}^n$, $|x| \geq r + 1$, relation $\langle \operatorname{grad} V(x), x - w \rangle \geq 0$ holds for all $w \in M_B(x)$.

So, we can study problem (2.26) with the assumption that

$(M)'$ $\langle \operatorname{grad} V(x), x - w \rangle \geq 0$ for all $w \in M(x)$ provided $|x| \geq r + 1$.

Substitute existence problem (2.26) with the problem of the existence of $x \in \mathbf{R}^n$ such that

$$0 \in \theta_T \circ \Pi_F(x) - M(x),$$

where the map θ_T and the multimap Π_F are defined as in (2.31) and (2.30), respectively.

Let $M = (\Sigma_q \circ \cdots \circ \Sigma_1) \in J^c(\mathbf{R}^n, \mathbf{R}^n)$, where $\Sigma_i \in J(X_{i-1}, X_i), i = 1, \cdots, q$; $X_0 = X_q = \mathbf{R}^n$, and X_i are normed spaces for all $0 < i < q$.

Define the following maps and multimaps:

$$\tilde{\Sigma}_1: \mathbf{R}^n \to K(\mathbf{R}^n \times X_1), \ \tilde{\Sigma}_1(x) = \{x\} \times \Sigma_1(x),$$

$$\tilde{\Sigma}_i: \mathbf{R}^n \times X_{i-1} \to K(\mathbf{R}^n \times X_i), \ \tilde{\Sigma}_i(x, y) = \{x\} \times \Sigma_i(y), \ \forall i = 2, \cdots, q.$$

$$\tilde{M}: \mathbf{R}^n \to K(\mathbf{R}^n \times \mathbf{R}^n), \ \tilde{M}(x) = \{x\} \times M(x),$$

$$\tilde{\Pi}_F: \mathbf{R}^n \times \mathbf{R}^n \to K(C([0, T]; \mathbf{R}^n) \times \mathbf{R}^n), \ \tilde{\Pi}_F(x, y) = \{\Pi_F(x)\} \times \{y\},$$

$$\tilde{\theta}_T: C([0, T]; \mathbf{R}^n) \times \mathbf{R}^n \to \mathbf{R}^n \times \mathbf{R}^n, \ \tilde{\theta}_T(u, y) = \{\theta_T(u)\} \times \{y\},$$

$$f: \mathbf{R}^n \times \mathbf{R}^n \to \mathbf{R}^n, \ f(x, y) = x - y.$$

It is clear that $\tilde{\Sigma}_i$ ($1 \leq i \leq q$) and $\tilde{\Pi}_F$ are J-multimaps; $\tilde{\theta}_T$, f are continuous maps and

$$\theta_T \circ \Pi_F - M = f \circ \tilde{\theta}_T \circ \tilde{\Pi}_F \circ \tilde{M} = f \circ \tilde{\theta}_T \circ \tilde{\Pi}_F \circ \tilde{\Sigma}_q \circ \cdots \circ \tilde{\Sigma}_1.$$

Therefore, $\theta_T \circ \Pi_F - M: \mathbf{R}^n \to K(\mathbf{R}^n)$ is a J^c-multimap.

Define multimaps

$$A: \mathbf{R}^n \to P(\mathbf{R}^n), \ A(x) = \{y \in \mathbf{R}^n : \langle y, \varphi(x) \operatorname{grad} V(x) \rangle \geq 0\},$$

and
$$F_A(t, x) = F(t, x) \cap A(x), \quad (t, x) \in [0, T] \times \mathbf{R}^n ,$$
where $\varphi(x)$ is defined above.

It is easy to verify that F_A is an L^1-upper Carathéodory multimap with α-sublinear growth and

$$\langle grad\, V(x), y \rangle \geq 0, \text{ for all } y \in F_A(t, x) \text{ provided } |x| > r.$$

Following Lemmas 2.3 and 2.7 we can choose a sufficiently large number $R > r + T + 1$ such that for every $(\lambda, x) \in [0, 1] \times \mathbf{R}^n$, $|x| = R$, we have:

(a) every solution $u: [0, T] \to \mathbf{R}^n$ of the Cauchy problem:

$$\begin{cases} u'(t) \in \Psi(t, u(t), \lambda) = \lambda W_V(u(t)) + (1 - \lambda) F_A(t, u(t)), \\ u(0) = x, \end{cases} \quad (2.32)$$

satisfies the condition: $|u(t)| > r, \; \forall t \in [0, T]$;
(b) there is $t_r \in (0, T]$ such that for all $u \in \Pi_{W_V}(x)$

$$\langle u(t) - x, grad\, V(x) \rangle > 0, \quad \forall t \in (0, t_r],$$

where Π_{W_V} is defined analogously to Π_F.

Now set $x \in \partial B_{\mathbf{R}^n}(0, R)$ and $z \in \theta_{t_r} \circ \Pi_{W_V}(x) - M(x)$. Then there exist $u \in \Pi_{W_V}(x)$ and $w \in M(x)$ such that $z = u(t_r) - w$. From the choice of R it follows that
$$\langle x - w, grad\, V(x) \rangle \geq 0 \text{ for all } w \in M(x).$$
Hence,

$$\langle z, grad\, V(x) \rangle = \langle u(t_r) - x, grad\, V(x) \rangle + \langle x - w, grad\, V(x) \rangle > 0.$$

Therefore, the vector fields $\theta_{t_r} \circ \Pi_{W_V} - M$ and $grad\, V$ are homotopic on $\partial B_{\mathbf{R}^n}(0, R)$. So,
$$deg(\theta_{t_r} \circ \Pi_{W_V} - M, B_{\mathbf{R}^n}(0, R)) = ind\, V.$$

If $0 \in \theta_T \circ \Pi_{F_A}(x) - M(x)$ for some $x \in \partial B_{\mathbf{R}^n}(0, R)$, then the theorem is proved, otherwise consider the following multimap

$$\Sigma: B_{\mathbf{R}^n}(0, R) \times [0, 1] \to K(\mathbf{R}^n),$$

$$\Sigma(x, \lambda) = \theta_{\lambda t_r + (1-\lambda)T} \circ \Pi_{\Psi_\lambda}(x) - M(x),$$

where $\Psi_\lambda(t, x) = \Psi(t, x, \lambda)$.

It is easy to verify that Σ is a J^c-multimap. Assume that there is $(x, \lambda) \in \partial B_{\mathbf{R}^n}(0, R) \times (0, 1]$ such that

$$0 \in \Sigma(x, \lambda).$$

Then there is a solution $u(\cdot)$ of problem (2.32) and $w \in M(x)$ such that

$$u(\lambda t_r + (1 - \lambda)T) = w.$$

From (a) it follows that $|u(t)| > r \geq r_V$ for all $t \in [0, T]$. Therefore,

$$grad\, V(u(t)) \neq 0, \; \forall t \in [0, T].$$

Hence,

$$\langle \lambda W_V(u(s)) + (1 - \lambda)y, grad\, V(u(s)) \rangle > 0$$

for all $s \in [0, T]$ and all $y \in F_A(s, u(s))$. So,

$$V\left(u(\lambda t_r + (1 - \lambda)T)\right) - V(x) = \int_0^{\lambda t_r + (1-\lambda)T} \langle u'(s), grad\, V(u(s)) \rangle ds > 0.$$

Consequently, $V(w) > V(x)$, that is a contradiction.

Thus, Σ is a homotopy connecting multimaps

$$\theta_{t_r} \circ \Pi_{W_V} - M \text{ and } \theta_T \circ \Pi_{F_A} - M.$$

Therefore,

$$deg(\theta_T \circ \Pi_{F_A} - M, B_{\mathbf{R}^n}(0, R)) = ind\, V \neq 0.$$

So, problem (2.26) has a solution.   □

Corollary 2.5. *Let conditions $(A1)$–$(A2)$ hold. Assume that there is $r > 0$ such that for every $(t, x) \in [0, T] \times \mathbf{R}^n$, $|x| \geq r$, the following relations hold:*

a) $\langle x, y \rangle > 0$ for at least one point $y \in F(t, x)$;
b) $\|M(x)\| = \max\{|w|: w \in M(x)\} \leq |x|$.

Then problem (2.26) has a solution.

Corollary 2.6 (Existence of anti-periodic solutions).   *Let conditions $(A1)$–$(A2)$ hold. In addition, assume that there exists $r > 0$ such that for every $(t, x) \in I \times \mathbf{R}^n$, $|x| \geq r$, there is at least one point $y \in F(t, x)$ such that*

$$\langle x, y \rangle \geq 0.$$

Then the anti-periodic problem

$$\begin{cases} u'(t) \subset F(t, u(t)), & \text{for a.e. } t \in [0, T], \\ u(T) = -u(0), \end{cases} \tag{2.33}$$

has a solution.

Proof. The conclusions of the Corollaries 2.5 and 2.6 follow immediately from Theorem 2.10 by using the guiding function $V: \mathbf{R}^n \to \mathbf{R}$, $V(x) = \frac{1}{2}|x|^2$.   □

Let us mention that the necessity of studying the existence of anti-periodic solutions for differential equations and inclusions arises in the investigation of many problems of physics (see, e.g., [118, 122, 127]), wavelet theory (see, e.g., [33]) and others branches of contemporary science. Some existence theorems for anti-periodic solutions are presented in [3, 35, 36].

Example 2.1. Consider the following problem

$$\begin{cases} u'(t) \in F(u(t)), \text{ for a.e. } t \in [0, 1], \\ u(1) \in \left[\frac{1}{2}u(0) + 1, \frac{1}{2}u(0) + 2\right], \end{cases} \tag{2.34}$$

where multimap $F: \mathbf{R} \to Kv(\mathbf{R})$ is defined by

$$F(x) = \begin{cases} 1 & \text{if } x > 0, \\ -1 & \text{if } x < 0, \end{cases}$$

and $F(0) = [-1, 1]$.

In this situation, $M(x) = \left[\frac{1}{2}x + 1, \frac{1}{2}x + 2\right]$, $x \in \mathbf{R}$. It is clear that all conditions in Corollary 2.5 hold. So, problem (2.34) has a solution.

2.5   Global Bifurcation Problems

The existence of a branch of non-trivial solutions of an operator-equation from a bifurcation point was studied first by M.A. Krasnosel'skii [89]. The global bifurcation theorem for single-valued case was proved by P. Rabinowitz [124]. The bifurcation problem for inclusions with convex-valued multimaps was studied by J.C. Alexander and P.M. Fitzpatrick [4]. The authors of this work gave sufficient conditions under which the set of all non-trivial solutions near the point $(0, 0)$ admits either a bifurcation to infinity, a bifurcation to the border of the considered domain, or a bifurcation to some trivial solution of the inclusion. After this work, in the other studies the bifurcation theory for inclusions was extended to the case when multimap takes non-convex values (see, e.g., [64, 66, 96]).

Recently, bifurcation theory has been also extended to the case of linear Fredholm inclusions (see, e.g., [59, 60, 103]). Some other results on the bifurcation theory for inclusions and differential inclusions of various types can be found, e.g., in [42, 43, 48–52, 69, 81, 99, 101, 102, 104, 115, 129] and others.

The application of topological tools is the major method for studying bifurcation problems. A *global bifurcation index* at a given point is evaluated by using topological degrees. If the global bifurcation index is non-zero, then the global structure of solutions of the considering problem can be described. However, in practice this evaluation faces several difficulties due to the problem of handling techniques related to the topological degree in functional spaces.

It turns out that the MGF can be effectively applied to the evaluation of the bifurcation index. It should be mentioned that, as far as our knowledge, the first attempt in such direction was made by W. Kryszewski [96].

In this section, we present an approach of the MGF for the evaluation of the global bifurcation index and its application to study the global structure of periodic solutions of ordinary differential inclusions in finite-dimensional spaces.

2.5.1 Abstract Result

In this section, we present the application of the bifurcation index to the description of the global structure of branches of non-trivial solutions to a family of inclusions.

Let X be a Banach space. Consider the following one-parameter family of inclusions

$$x \in \mathscr{F}(x, \mu), \tag{2.35}$$

where $\mathscr{F}: X \times \mathbf{R} \to Kv(X)$ is a multimap.

Assume that:

$(\mathscr{F}1)$ $\mathscr{F}$ is completely upper semicontinuous and $0 \in \mathscr{F}(0, \mu)$ for all $\mu \in \mathbf{R}$;

$(\mathscr{F}2)$ for each μ, $0 < |\mu - \mu_0| \le \varepsilon_0$, there is $\delta_\mu > 0$ such that $x \notin \mathscr{F}(x, \mu)$ when $0 < \|x\| \le \delta_\mu$, where μ_0, ε_0 are given numbers.

Definition 2.17. A point $(0, \mu_*)$ is said to be a bifurcation point of inclusion (2.35) if for every open subset $U \subset X \times \mathbf{R}$ containing $(0, \mu_*)$ there exists a point $(x, \mu) \in U$ such that $x \ne 0$ and $x \in \mathscr{F}(x, \mu)$.

From $(\mathscr{F}1)$–$(\mathscr{F}2)$ it follows that for each μ, $0 < |\mu - \mu_0| \le \varepsilon_0$ the topological degree

$$deg\big(i - \mathscr{F}(\cdot, \mu), B_X(0, \delta_\mu)\big)$$

is well defined. Then the *bifurcation index of the multimap* $\mathscr{F}$ at $(0, \mu_0)$ can be defined as

$$Bi\big[\mathscr{F}; (0, \mu_0)\big] = \lim_{\mu \to \mu_0+} deg\big(i - \mathscr{F}(\cdot, \mu), B_X(0, \delta_\mu)\big)$$

$$- \lim_{\mu \to \mu_0^-} deg\big(i - \mathscr{F}(\cdot, \mu), B_X(0, \delta_\mu)\big).$$

Let us describe the geometric meaning of the bifurcation index. For each sufficiently small $\varepsilon \in (0, \varepsilon_0]$, where ε_0 is the constant in $(\mathscr{F}2)$, consider the multifield

$$F_r: \overline{U}_r \to Kv(X \times \mathbf{R}),$$

$$F_r(x, \mu) = \{x - \mathscr{F}(x, \mu), \|x\|^2 - r^2\},$$

where $r \in \left(0, \min\{\delta_{\mu_0-\varepsilon}, \delta_{\mu_0+\varepsilon}\}\right)$ is taken small enough and

$$U_r = \{(x, \mu) \in X \times \mathbf{R}: \|x\|^2 + (\mu - \mu_0)^2 < r^2 + \varepsilon^2\}.$$

It is clear that the vector field F_r is completely upper semicontinuous.
Let us mention that it has no zeros on the boundary ∂U_r.
  Indeed assume, to the contrary, that there is $(x, \mu) \in \partial U_r$ such that

$$0 \in F_r(x, \mu).$$

Then we obtain

$$\begin{cases} \|x\| = r, \\ x \in \mathscr{F}(x, \mu). \end{cases}$$

From $(x, \mu) \in \partial U_r$ it follows that $\mu = \mu_0 \pm \varepsilon$. From $(\mathscr{F}2)$ and the choice of r we obtain a contradiction.
So the topological degree $deg(F_r, \overline{U}_r)$ is well-defined and does not depend on the choice of r.
  The following statement is the generalization of the Ize's lemma (for more details we refer reader to [77, 78, 113]).

Lemma 2.8 (see [42, 69]). *For each sufficiently small* $\varepsilon \in (0, \varepsilon_0]$:

$$deg(F_r, \overline{U}_r) = -Bi[\mathscr{F}; (0, \mu_0)] .$$

Let us denote by $\mathscr{S}$ the set of all non-trivial solutions to inclusion (2.35), i.e.,

$$\mathscr{S} = \{(x, \mu) \in X \times \mathbf{R}: x \neq 0 \text{ and } x \in \mathscr{F}(x, \mu)\}.$$

The following assertion follows easily from [59, 96].

Theorem 2.11. *Under conditions* $(\mathscr{F}1)$–$(\mathscr{F}2)$, *assume that* $Bi\left[\mathscr{F}; (0, \mu_0)\right] \neq 0$. *Then there exists a connected subset* $\mathscr{R} \subset \mathscr{S}$ *such that* $(0, \mu_0) \in \overline{\mathscr{R}}$ *and one of the following cases occurs:*

(a) $\mathscr{R}$ *is unbounded;*
(b) $(0, \mu_*) \in \overline{\mathscr{R}}$ *for some* $\mu_* \neq \mu_0$.

2.5.2 Global Bifurcation of Periodic Solutions

Here we want to use the MGF for the evaluation of the bifurcation index for a family of differential inclusions in a finite-dimensional space. Then, the abstract result of the previous section is applied to the study of the global bifurcation of periodic solutions for this family.

The Setting of the Problem

Let $I = [0, T]$. We consider the following family of inclusions

$$x'(t) \in F(t, x(t), \mu) \text{ for a.e. } t \in I, \tag{2.36}$$

$$x(0) = x(T). \tag{2.37}$$

We assume the following conditions:

($H1$) The multimap $F: \mathbf{R} \times \mathbf{R}^n \times \mathbf{R} \to Kv(\mathbf{R}^n)$ is T-periodic L^p-upper Carathéodory, $p \geq 1$.

($H2$) The multimap $F(0, \cdot, \cdot): \mathbf{R}^n \times \mathbf{R} \to Kv(\mathbf{R}^n)$ is u.s.c.

($H3$) $0 \in F(s, 0, \mu)$ for all $\mu \in \mathbf{R}$ and a.e. $s \in [0, T]$.

We know (see Sect. 1.1.2) that under condition ($H1$) the superposition multioperator $\mathscr{P}_F: C(I, \mathbf{R}^n) \times \mathbf{R} \to Cv(L^p(I, \mathbf{R}^n))$ is well-defined and closed.

By a T-periodic solution to problem (2.36)–(2.37) we mean a pair $(x, \mu) \in W_T^{1,p}(I, \mathbf{R}^n) \times \mathbf{R}$ satisfying (2.36). From ($H3$) it follows that $(0, \mu)$ is a solution to problem (2.36)–(2.37) for each $\mu \in \mathbf{R}$. These solutions are called *trivial*. Let us denote by $\mathscr{S}$ the set of all nontrivial solutions of problem (2.36)–(2.37).

Global Structure of $\mathscr{S}$ When $p = 1$

We consider the global structure of T-periodic solutions to problem (2.36)–(2.37) when $p = 1$. Notice that T-periodic solutions of problem (2.36)–(2.37) are fixed points of the following family of integral multioperators

$$J_T: C(I, \mathbf{R}^n) \times \mathbf{R} \to Kv(C(I, \mathbf{R}^n)),$$

$$J_T(x, \mu) = \left\{ u: u(t) = x(T) + \int_0^t f(s)\, ds, \ f \in \mathscr{P}_F(x, \mu) \right\}.$$

It is easy to see that the multioperator J_T is completely upper semicontinuous. Extending Definition 2.1, we say that for a fixed $\mu \in \mathbf{R}$, a point $x_0 \in \mathbf{R}^n$ is a T-non-recurrence point of trajectories of inclusion (2.36), if for every nontrivial solution x of inclusion (2.36) satisfying condition $x(0) = x_0$ we have $x(t) \neq x_0$ for all $t \in (0, T]$.

The following theorem is a basic tool for considering the application of the MGF to bifurcation problems.

Theorem 2.12. *Let conditions ($H1$)–($H3$) hold. Assume that for each μ with*

$$0 < |\mu - \mu_0| \leq \varepsilon_0, \text{ where } \mu_0, \varepsilon_0 \text{ are given numbers,}$$

the following conditions hold:

$(H4)$ *there exists a sufficiently small $\varepsilon_\mu > 0$ such that from the fact that (x, μ) is a non-trivial solution of inclusion (2.36) with the initial condition $x(0) = 0$, it follows that $\|x\|_C \geq \varepsilon_\mu$;*

$(H5)$ *there is $\delta_\mu \in (0, \varepsilon_\mu)$, where ε_μ is the constant in $(H4)$, such that every point $y \in B_{\mathbf{R}^n}(0, \delta_\mu) \setminus \{0\}$ is a T-non-recurrence point of trajectories of inclusion (2.36);*

$(H6)$ *multifield $Q_\mu : \mathbf{R}^n \to Kv(\mathbf{R}^n)$,*

$$Q_\mu(y) = -F(0, y, \mu),$$

has no zeros on $B_{\mathbf{R}^n}(0, \delta_\mu) \setminus \{0\}$.

Then $x \notin J_T(x, \mu)$ provided $0 < \|x\|_C \leq \delta_\mu$ and

$$\deg(i - J_T(\cdot, \mu), B_C(0, \delta_\mu)) = \deg(Q_\mu, B_{\mathbf{R}^n}(0, \delta_\mu)).$$

Proof. Fixing $\mu, 0 < |\mu - \mu_0| \leq \varepsilon_0$, we consider the following family of multimaps

$$F_\mu(t, y, \lambda) = F(\lambda t, y, \mu), \quad \lambda \in [0, 1]$$

and the corresponding family of multifields

$$\Psi_\mu : C(I, \mathbf{R}^n) \times [0, 1] \to Kv(C(I, \mathbf{R}^n))$$

$$\Psi_\mu(x, \lambda) = \left\{ u : u(t) \right.$$

$$= x(t) - x(T) - \lambda \int_0^t f(s)\,ds - (1 - \lambda) \int_0^T f(s)\,ds, \ f \in \mathscr{P}_{F_\mu}(x, \lambda) \right\}.$$

It is clear that the family of multifields Ψ_μ corresponds to the completely u.s.c. family of multimaps $i - \Psi_\mu$.

Let us show that Ψ_μ has no singular points on $(B_C(0, \delta_\mu) \setminus \{0\}) \times [0, 1]$.

To the contrary, assume that there is

$$(x_*, \lambda_*) \in (B_C(0, \delta_\mu) \setminus \{0\}) \times [0, 1],$$

such that $0 \in \Psi_\mu(x_*, \lambda_*)$. It means that there is a function $f \in L^1(I, \mathbf{R}^n)$ such that

$$f(s) \in F(\lambda_* s, x_*(s), \mu) \text{ for a.e. } s \in I,$$

and

$$x_*(t) = x_*(T) + \lambda_* \int_0^t f(s)\,ds + (1 - \lambda_*) \int_0^T f(s)\,ds, \qquad (2.38)$$

for all $t \in [0, T]$.

For $t = 0$ we have

$$x_*(0) = x_*(T) + (1 - \lambda_*) \int_0^T f(s)ds,$$

Taking $t = T$, we obtain

$$\int_0^T f(s)ds = 0. \tag{2.39}$$

Therefore, $x_*(0) = x_*(T)$.

From (2.38) it follows that

$$x_*'(t) = \lambda_* f(t) \in \lambda_* F(\lambda_* t, x_*(t), \mu)$$

for a.e. $t \in [0, T]$.

Thus, x_* is a solution of the inclusion

$$x'(t) \in \lambda_* F(\lambda_*, x(t), \mu).$$

(i) If $\lambda_* = 0$, then $x_*(t) = x_0 \in B_{\mathbf{R}^n}(0, \delta_\mu) \setminus \{0\}$ for all $t \in [0, T]$. We have

$$\int_0^T F(0, x_0, \mu)ds = T \cdot F(0, x_0, \mu).$$

From (2.39) and

$$\int_0^T f(s)ds \in \int_0^T F(0, x_0, \mu)ds$$

it follows that

$$0 \in F(0, x_0, \mu),$$

that is a contradiction because of the fact that Q_μ has no zeros on $B_{\mathbf{R}^n}(0, \delta_\mu) \setminus \{0\}$.

(ii) Let $\lambda_* \neq 0$. Consider a function $z_*(t) = x_*(\frac{t}{\lambda_*})$. Then for a.e. $t \in [0, \lambda_* T]$ we have

$$z_*'(t) = \frac{1}{\lambda_*} x_*'(\frac{t}{\lambda_*}) = f(\frac{t}{\lambda_*}) \in F(t, x_*(\frac{t}{\lambda_*}), \mu) = F(t, z_*(t), \mu).$$

Thus, (z_*, μ) is a solution to inclusion (2.36) on the interval $[0, \lambda_* T]$.

The case $x_*(0) = 0$: in this situation the pair $(\tilde{z}_*, \mu)$, where

$$\tilde{z}_*(t) = \begin{cases} z_*(t) & if \ t \in [0, \lambda_* T] \\ 0 & if \ t \in [\lambda_* T, T] \end{cases}$$

is a solution to inclusion (2.36) with the initial condition $\tilde{z}_*(0) = 0$. On the other hand,

$$\|\tilde{z}_*\|_C \leq \delta_\mu < \varepsilon_\mu$$

giving a contradiction.

The case $x_*(0) \neq 0$: w.l.o.g. we can assume that z_* is extended to $[0, T]$. We have $z_*(0) = x_*(0) = x_*(T) = z_*(\lambda_* T)$. Hence, inclusion (2.36) has a nontrivial solution (z_*, μ) such that $z_*(0) \in B_{\mathbf{R}^n}(0, \delta_\mu) \setminus \{0\}$ and $z_*(0) = z_*(\lambda_* T)$, that is a contradiction with the T-non-recurrence of the trajectories of inclusion (2.36).

Thus, Ψ_μ is a homotopy connecting multifields

$$\Psi_\mu^{(1)} = i - J_T(\cdot, \mu),$$

and

$$\Psi_\mu^{(0)} = i - \Gamma_\mu^{(0)}(\cdot),$$

where multioperator $\Gamma_\mu^{(0)}: B_C(0, \delta_\mu) \to Kv(C(I, \mathbf{R}^n))$ is defined as

$$\Gamma_\mu^{(0)}(x) = x(T) + \int_0^T F(0, x(s), \mu)ds.$$

This multioperator has its range in the space $C_{[0,T]}^n$ of constant functions which can be identified with $\mathbf{R}^n$. Then

$$deg(\Psi_\mu^{(0)}, B_C(0, \delta_\mu)) = deg(\hat{\Psi}_\mu^{(0)}, B_{\mathbf{R}^n}(0, \delta_\mu)),$$

where $\hat{\Psi}_\mu^{(0)} = \Psi_\mu^{(0)}|_{\mathbf{R}^n}$ is defined by

$$\hat{\Psi}_\mu^{(0)}(y) = -\int_0^T F(0, y, \mu)ds = -T \cdot F(0, y, \mu).$$

So we obtain

$$deg(i - J_T(\cdot, \mu), B_C(0, \delta_\mu)) = deg(Q_\mu, B_{\mathbf{R}^n}(0, \delta_\mu)). \qquad \square$$

Definition 2.18. A continuously differentiable function $V: \mathbf{R}^n \to \mathbf{R}$ is said to be a local non-degenerate potential if there exists a sufficiently small number $r > 0$ such that the gradient

$$grad\, V(x) = \left(\frac{\partial V(x)}{\partial x_1}, \frac{\partial V(x)}{\partial x_2}, \cdots, \frac{\partial V(x)}{\partial x_n} \right)$$

is not equal zero provided $0 < |x| \leq r$.

It is clear that the topological degree

$$deg(grad\, V,\, B_{\mathbf{R}^n}(0, r'))$$

is well-defined and does not depend on $r' \in (0, r)$. This number is called *local index of a non-degenerate potential V* and is denoted by *ind V*.

Definition 2.19. For each $\mu \in \mathbf{R}$, a continuously differentiable function $V_\mu : \mathbf{R}^n \to \mathbf{R}$ is said to be a local guiding function for inclusion (2.36), if there exists a sufficiently small number $\tau_\mu > 0$ such that for every $y \in F(t, x, \mu)$:

$$\begin{cases} \langle grad\, V_\mu(x), y \rangle > 0 \, for \; t = 0 \; and \; a.e. \; t \in (0, \tau_\mu), \; 0 < |x| < \tau_\mu, \\ \langle grad\, V_\mu(x), y \rangle \geq 0 \, for \; a.e. \; t \in [\tau_\mu, T]. \end{cases}$$

From Definition 2.19 it follows that if V_μ is a local guiding function for inclusion (2.36) then V_μ is a non-degenerate potential and vector fields $-grad\, V_\mu$ and Q_μ are homotopic on $\partial B_{\mathbf{R}^n}(0, r)$ for every $0 < r < \tau_\mu$. Therefore

$$deg(Q_\mu,\, B_{\mathbf{R}^n}(0, r)) = deg(-grad\, V_\mu,\, B_{\mathbf{R}^n}(0, r)) = (-1)^n ind\, V_\mu.$$

Theorem 2.13. *Let conditions $(H1)$–$(H4)$ hold for $p = 1$. Assume that for each μ,*

$$0 < |\mu - \mu_0| \leq \varepsilon_0, \; where \; \varepsilon_0 \; and \; \mu_0 \; are \; given \; numbers,$$

there is a local guiding function V_μ for inclusion (2.36) such that

$$\lim_{\mu \to \mu_0^+} ind\, V_\mu - \lim_{\mu \to \mu_0^-} ind\, V_\mu \neq 0.$$

Then there exists a connected subset $\mathcal{W} \subset \mathcal{S}$ such that $(0, \mu_0) \in \overline{\mathcal{W}}$ and either $\mathcal{W}$ is unbounded, or $(0, \mu_*) \in \overline{\mathcal{W}}$ for some $\mu_* \neq \mu_0$.

Proof. Let us show that the multioperator J_T satisfies all conditions in Theorem 2.11.

In fact, condition $(\mathscr{F}1)$ can be easily verified. In order to verify condition $(\mathscr{F}2)$ and calculate the bifurcation index $Bi[J_T; (0, \mu_0)]$ we fix μ, $0 < |\mu - \mu_0| \leq \varepsilon_0$, and choose

$$0 < \delta_\mu < \min\{\varepsilon_\mu, \tau_\mu\},$$

where $\varepsilon_\mu, \tau_\mu$ are numbers in $(H4)$ and Definition 2.19, respectively.

Let us show that $B_{\mathbf{R}^n}(0, \delta_\mu) \setminus \{0\}$ is the set consisting of T-non-recurrence points of trajectories of inclusion (2.36). Indeed, take $x_0 \in B_{\mathbf{R}^n}(0, \delta_\mu) \setminus \{0\}$ and let x be an arbitrary nontrivial solution of inclusion (2.36) with initial condition $x(0) = x_0$. Assume that there is $t_* \in (0, T]$ such that $x(t_*) = x(0)$. Since $|x_0| < \tau_\mu$, there exists $t_\mu \in (0, \tau_\mu)$ such that $t_\mu < t_*$ and $|x(t)| < \tau_\mu$ for all $t \in (0, t_\mu)$. Therefore

$$0 = V_\mu(x(t_*)) - V_\mu(x(0)) = \int_0^{t_*} \langle grad\, V_\mu(x(s)), x'(s) \rangle\, ds$$

$$= \int_0^{t_\mu} \langle grad\, V_\mu(x(s)), x'(s) \rangle\, ds + \int_{t_\mu}^{t_*} \langle grad\, V_\mu(x(s)), x'(s) \rangle\, ds > 0,$$

giving the contradiction.

Notice that for every μ, $0 < |\mu - \mu_0| \le \varepsilon_0$, from the existence of the guiding function V_μ for inclusion (2.36) it follows that the vector field $Q_\mu = -F(0, y, \mu)$ has no zeros on $B_{\mathbf{R}^n}(0, \delta_\mu) \setminus \{0\}$. By Theorem 2.12 we have that $x \notin J_T(x, \mu)$ provided $0 < \|x\|_C \le \delta_\mu$ and

$$\lim_{\mu \to \mu_0^+} deg(i - J_T(\cdot, \mu), B_C(0, \delta_\mu)) - \lim_{\mu \to \mu_0^-} deg(i - J_T(\cdot, \mu), B_C(0, \delta_\mu))$$

$$= \lim_{\mu \to \mu_0^+} deg(Q_\mu, B_{\mathbf{R}^n}(0, \delta_\mu)) - \lim_{\mu \to \mu_0^-} deg(Q_\mu, B_{\mathbf{R}^n}(0, \delta_\mu))$$

$$= (-1)^n \Big(\lim_{\mu \to \mu_0^+} ind\, V_\mu - \lim_{\mu \to \mu_0^-} ind\, V_\mu \Big).$$

Hence,

$$Bi[J_T; (0, \mu_0)] = (-1)^n \Big(\lim_{\mu \to \mu_0^+} ind\, V_\mu - \lim_{\mu \to \mu_0^-} ind\, V_\mu \Big) \ne 0.$$

To conclude the proof we need only to apply Theorem 2.11.   □

Global Bifurcation When $p = 2$

Now, by introducing the notion of local integral guiding functions for inclusion (2.36) we consider the global structure of the set of all T-periodic solutions of problem (2.36)–(2.37). Assume that F is a T-periodic L^2-upper Carathéodory multimap satisfying condition $(H3)$.

Define the operator $\ell \colon W_T^{1,2}(I, \mathbf{R}^n) \to L^2(I, \mathbf{R}^n)$ as

$$\ell x = x'.$$

It is clear that ℓ is a linear Fredholm operator of index zero and

$$Ker\,\ell \cong \mathbf{R}^n \cong Coker\,\ell.$$

Then we can substitute problem (2.36)–(2.37) by the following family of operator inclusions

$$\ell x \in \mathscr{P}_F(x, \mu),$$

or by equivalently (see, Sect. 1.3)

$$x \in G(x, \mu), \tag{2.40}$$

where

$$G: C_T(I, \mathbf{R}^n) \times \mathbf{R} \to Kv(C_T(I, \mathbf{R}^n)),$$

$$G(x, \mu) = Px + (\Lambda\Pi + K_{P,Q})\mathscr{P}_F(x, \mu). \tag{2.41}$$

Recall that

$$\Pi: L^2(I, \mathbf{R}^n) \to \mathbf{R}^n,$$

is defined by

$$\Pi f = \frac{1}{T} \int_0^T f(s)\, ds$$

and the homomorphism $\Lambda: \mathbf{R}^n \to \mathbf{R}^n$ can be treated as the identity map.

Definition 2.20. For each $\mu \in \mathbf{R}$, a continuously differentiable function $V_\mu: \mathbf{R}^n \to \mathbf{R}$ is said to be a local integral guiding function for inclusion (2.36), if there exists a sufficiently small number $\pi_\mu > 0$ such that from $x \in W_T^{1,2}(I, \mathbf{R}^n)$ with $0 < \|x\|_2 \leq \pi_\mu$ it follows that

$$\int_0^T \langle grad\, V_\mu(x(s)), f(s)\rangle\, ds > 0$$

for all $f \in \mathscr{P}_F(x, \mu)$.

Notice that the local integral guiding function V_μ is a non-degenerate potential. In fact, for every $y \in \mathbf{R}^n$ with $0 < |y| \leq \frac{\pi_\mu}{\sqrt{T}}$, considering y as a constant function we have

$$\int_0^T \langle grad\, V_\mu(y), f(s)\rangle\, ds = \langle grad\, V_\mu(y), \int_0^T f(s)ds\rangle = T\langle grad\, V_\mu(y), \Pi f\rangle > 0$$

for all $f \in \mathscr{P}_F(y, \mu)$.

Hence, $grad\, V_\mu(y) \neq 0$ provided $0 < |y| \leq \frac{\pi_\mu}{\sqrt{T}}$.

Theorem 2.14. *Let F be a T-periodic L^2-Carathéodory multimap satisfying conditions (H3). Assume that for each μ, $0 < |\mu - \mu_0| \leq \varepsilon_0$, where ε_0, μ_0 are given numbers, there exists a local integral guiding function V_μ for inclusion (2.36) such that*

$$\lim_{\mu \to \mu_0^+} ind\, V_\mu - \lim_{\mu \to \mu_0^-} ind\, V_\mu \neq 0.$$

Then there is a connected subset $\mathscr{W} \subset \mathscr{S}$ such that $(0, \mu_0) \in \overline{\mathscr{W}}$ and either $\mathscr{W}$ is unbounded or $(0, \mu_*) \in \overline{\mathscr{R}}$ for some $\mu_* \neq \mu_0$.

Proof. We show that multioperator G defined in (2.41) satisfies all conditions in Theorem 2.11. At first, the space $L^2(I, \mathbf{R}^n)$ can be decomposed by

$$L^2(I, \mathbf{R}^n) = \mathscr{L}_0 \oplus \mathscr{L}_1,$$

where $\mathscr{L}_0 = Coker\, \ell$, $\mathscr{L}_1 = Im\, \ell$. The corresponding decomposition of an element $f \in L^2(I, \mathbf{R}^n)$ is denoted by

$$f = f_0 + f_1,\, f_0 \in \mathscr{L}_0,\, f_1 \in \mathscr{L}_1.$$

STEP 1. From $(H3)$ it follows that $0 \in G(0, \mu)$ for all $\mu \in \mathbf{R}$. Let

$$\Phi : C_T(I, \mathbf{R}^n) \times \mathbf{R} \times [0, 1] \to Cv(L^2(I, \mathbf{R}^n)),$$

$$\Phi(x, \mu, \lambda) = \chi(\mathscr{P}_F(x, \mu), \lambda),$$

where

$$\chi(f_0 + f_1, \lambda) = f_0 + \lambda f_1.$$

We prove that the multimap

$$\Sigma : C_T(I, \mathbf{R}^n) \times \mathbf{R} \times [0, 1] \to Kv(C_T(I, \mathbf{R}^n)),$$

$$\Sigma(x, \mu, \lambda) = Px + (\Lambda\Pi + K_{P,Q})\Phi(x, \mu, \lambda),$$

is completely u.s.c.
Indeed, from the fact that the multioperator $\mathscr{P}_F$ is closed and the operator $(\Lambda\Pi + K_{P,Q}) \circ \chi$ is linear and continuous it follows that the multimap $(\Lambda\Pi + K_{P,Q})\chi \circ \mathscr{P}_F$ is closed (see Proposition 1.5). Further, for every bounded subset $U \subset C_T(I, \mathbf{R}^n) \times \mathbf{R}$ the set $\mathscr{P}_F(U)$ is bounded in $L^2(I, \mathbf{R}^n)$. Then the set $(\Lambda\Pi + K_{P,Q})\chi \circ \mathscr{P}_F(U)$ is bounded in $W_T^{1,2}(I, \mathbf{R}^n)$ and by the compact embedding property (see, e.g. [14, 41]), the set $(\Lambda\Pi + K_{P,Q})\chi \circ \mathscr{P}_F(U)$ is relatively compact in $C_T(I, \mathbf{R}^n)$. Finally, our assertion follows from the fact that the operator P is continuous and takes values in a finite dimensional space. In particular, the multimap $G = \Sigma(\cdot, \cdot, 1)$ is completely u.s.c.. So condition $(\mathscr{F}1)$ holds.
STEP 2. For each μ, $0 < |\mu - \mu_0| \le \varepsilon_0$, choosing r_μ such that

$$0 < r_\mu \le \min\{\pi_\mu, \frac{\pi_\mu}{\sqrt{T}}\},$$

where π_μ is a constant from Definition 2.20, assume that (x, μ), $x \in B_C(0, r_\mu)$ is a nontrivial solution to inclusion (2.40). Then there is $f \in \mathscr{P}_F(x, \mu)$ such that $x'(t) = f(t)$ for a.e. $t \in [0, T]$.

Since $0 < \|x\|_2 \leq \pi_\mu$ we have

$$\int_0^T \langle grad\, V_\mu(x(s)),\, f(s) \rangle\, ds = \int_0^T \langle grad\, V_\mu(x(s)),\, x'(s) \rangle\, ds$$

$$= V_\mu(x(T)) - V_\mu(x(0)) > 0,$$

giving a contradiction, i.e., inclusion (2.40) has no nontrivial solutions on $B_C(0, r_\mu)$. Therefore $(\mathscr{F}2)$ holds.

STEP 3. Now we evaluate the bifurcation index $Bi\big[G;(0,\mu_0)\big]$. Toward this goal, we fix μ, $0 < |\mu - \mu_0| \leq \varepsilon_0$, and choose r_μ as in Step 2. Consider the following family of inclusions

$$x \in \Sigma_\mu(x, \lambda), \tag{2.42}$$

where $\Sigma_\mu: C_T(I, \mathbf{R}^n) \times [0, 1] \to Kv(C_T(I, \mathbf{R}^n))$,

$$\Sigma_\mu(x, \lambda) = Px + (\Lambda\Pi + K_{P,Q})\Phi(x, \mu, \lambda).$$

As in Step 1, multioperator Σ_μ is completely u.s.c.. Assume that there is a solution $(x^*, \lambda^*) \in \partial B_C(0, r_\mu) \times [0, 1]$ of inclusion (2.42). Then there exists a function $f^* \in \mathscr{P}_F(x^*, \mu)$ such that

$$x^* = Px^* + (\Lambda\Pi + K_{P,Q}) \circ \chi(f^*, \lambda^*),$$

or equivalently,

$$\begin{cases} \ell x^* = \lambda^* f_1^* \\ 0 = f_0^*, \end{cases}$$

where $f_0^* + f_1^* = f^*$, $f_0^* \in \mathscr{L}_0$ and $f_1^* \in \mathscr{L}_1$.

Since $0 < \|x^*\|_2 \leq \pi_\mu$ we have

$$\int_0^T \langle grad\, V_\mu(x^*(s)),\, f^*(s) \rangle\, ds > 0.$$

If $\lambda^* \neq 0$, then

$$\int_0^T \langle grad\, V_\mu(x^*(s)),\, f^*(s) \rangle\, ds = \int_0^T \langle grad\, V_\mu(x^*(s)),\, \frac{1}{\lambda^*} x^{*\prime}(s) \rangle\, ds$$

$$= \frac{1}{\lambda^*} \big(V_\mu(x^*(T)) - V_\mu(x^*(0)) \big) = 0,$$

that is a contradiction.

If $\lambda^* = 0$, then $\ell x^* = 0$, i.e., $x^* \equiv a \in \mathbf{R}^n$, $|a| = r_\mu$. For every $f \in \mathscr{P}_F(a, \mu)$ we have

$$\int_0^T \langle grad\, V_\mu(a),\, f(s) \rangle\, ds > 0.$$

On the other hand,

$$\int_0^T \langle grad\, V_\mu(a), f(s)\rangle\, ds = \left\langle grad\, V_\mu(a), \int_0^T f(s)\, ds\right\rangle = T\langle grad\, V_\mu(a), \Pi f\rangle.$$

Therefore,

$$T\langle grad\, V_\mu(a), \Pi f\rangle > 0. \qquad (2.43)$$

Hence, $\Pi f \neq 0$ for all $f \in \mathscr{P}_F(a, \mu)$, in particular, $\Pi f^* \neq 0$. But $\Pi f^* = \Pi f_0^* = 0$, giving a contradiction.

Thus, Σ_μ is a homotopy connecting multimaps $\Sigma_\mu(\cdot, 1) = G(\cdot, \mu)$ and $\Sigma_\mu(\cdot, 0) = P + \Pi \mathscr{P}_F(\cdot, \mu)$. From the homotopy invariance property of the topological degree it follows that

$$deg(i - G(\cdot, \mu), B_C(0, r_\mu)) = deg(i - P - \Pi \mathscr{P}_F(\cdot, \mu), B_C(0, r_\mu)).$$

Multimap $P + \Pi \mathscr{P}_F(\cdot, \mu)$ takes its values in $\mathbf{R}^n$, then

$$deg(i - P - \Pi \mathscr{P}_F(\cdot, \mu), B_C(0, r_\mu)) = deg(i - P - \Pi \mathscr{P}_F(\cdot, \mu), B_{\mathbf{R}^n}(0, r_\mu)),$$

In the space $\mathbf{R}^n$ multifield $i - P - \Pi \mathscr{P}_F(\cdot, \mu)$ has the form

$$i - P - \Pi \mathscr{P}_F(\cdot, \mu) = -\Pi \mathscr{P}_F(\cdot, \mu),$$

therefore

$$deg(i - P - \Pi \mathscr{P}_F(\cdot, \mu), B_{\mathbf{R}^n}(0, r_\mu)) = deg(-\Pi \mathscr{P}_F(\cdot, \mu), B_{\mathbf{R}^n}(0, r_\mu)).$$

From (2.43) and Lemma 1.5 it follows that the vector fields $\Pi \mathscr{P}_F(\cdot, \mu)$ and $grad\, V_\mu$ are homotopic on $\partial B_{\mathbf{R}^n}(0, r_\mu)$, so

$$deg(-\Pi \mathscr{P}_F(\cdot, \mu), B_{\mathbf{R}^n}(0, r_\mu)) = deg(-grad\, V_\mu, B_{\mathbf{R}^n}(0, r_\mu)) = (-1)^n ind\, V_\mu.$$

Consequently,

$$\lim_{\mu \to \mu_0^+} deg(i - G(\cdot, \mu), B_C(0, r_\mu)) - \lim_{\mu \to \mu_0^-} deg(i - G(\cdot, \mu), B_C(0, r_\mu))$$

$$= \lim_{\mu \to \mu_0^+} deg(-\Pi \mathscr{P}_F(\cdot, \mu), B_{\mathbf{R}^n}(0, r_\mu)) - \lim_{\mu \to \mu_0^-} deg(-\Pi \mathscr{P}_F(\cdot, \mu), B_{\mathbf{R}^n}(0, r_\mu))$$

$$= (-1)^n \left(\lim_{\mu \to \mu_0^+} ind\, V_\mu - \lim_{\mu \to \mu_0^-} ind\, V_\mu \right).$$

Hence,

$$Bi[G;(0,\mu_0)] = (-1)^n \Big(\lim_{\mu \to \mu_0^+} ind\, V_\mu - \lim_{\mu \to \mu_0^-} ind\, V_\mu \Big) \neq 0.$$

To conclude the proof we need only to apply Theorem 2.11.   □

2.5.3   Application 1: Differential Inclusion with a Bounded Nonlinearity

Consider the following differential inclusion

$$x'(t) \in \mu x(t)\Big(a + F(t, x(t)) \Big), \tag{2.44}$$

where $F: \mathbf{R} \times \mathbf{R} \to Kv(\mathbf{R})$ is a T-periodic upper Carathéodory multimap; $a > 0$, $\mu \in \mathbf{R}$.

Denote by $\mathscr{S}$ the set of all non-trivial T-periodic solutions of inclusion (2.44).

Theorem 2.15. *Assume that:*

(A) there is $0 < K < a$ such that

$$\|F(t, y)\| = \max\{|z| : z \in F(t, y)\} < K$$

for all $y \in \mathbf{R}$ and a.e. $t \in [0, T]$.

Then there is a connected subset $\mathscr{W} \subset \mathscr{S}$ such that $(0, 0) \in \overline{\mathscr{W}}$ and $\mathscr{W}$ is unbounded.

Proof. It is clear that $(0, \mu)$ is a solution of inclusion (2.44) for every $\mu \in \mathbf{R}$ and $(y, 0)$ is a solution of inclusion (2.44) for every constant function $y \in \mathbf{R}$. Therefore $(0, 0)$ is a bifurcation point. Now let us show that $(0, 0)$ is the unique bifurcation point of (2.44).

To this aim we define a multimap $\tilde{F}: \mathbf{R} \times \mathbf{R} \times \mathbf{R} \to Kv(\mathbf{R})$ by

$$\tilde{F}(t, y, \mu) = \mu y \big(a + F(t, y) \big).$$

It is easy to see that $\tilde{F}$ is a T-periodic L^2-upper Carathéodory multimap.

We want to show that for every $\mu \neq 0$ the function

$$V_\mu : \mathbf{R} \to \mathbf{R},$$

$$V_\mu(y) = \frac{1}{2}\,\mu y^2,$$

is a local integral guiding function for inclusion (2.44).

In fact, let $x \in W_T^{1,2}([0, T]; \mathbf{R})$ and choose an arbitrary $f \in \mathscr{P}_F(x)$. Then

$$\tilde{f} = \mu x(a + f) \in \mathscr{P}_{\tilde{F}}(x, \mu).$$

We have

$$\int_0^T \Big\langle grad\, V_\mu(x(s)), \tilde{f}(s) \Big\rangle ds = \int_0^T \Big\langle \mu x(s), \mu a x(s) + \mu x(s) f(s) \Big\rangle ds$$

$$\geq \mu^2 \Big(a \|x\|_2^2 + \int_0^T x^2(s) f(s) ds \Big)$$

$$\geq \mu^2 \Big(a \|x\|_2^2 - \|x\|_2 \|xf\|_2 \Big)$$

$$\geq \mu^2 \|x\|_2^2 (a - K) > 0, \qquad (2.45)$$

for $\|x\|_2 > 0$.

Therefore, inclusion (2.44) has no non-trivial solution provided $\mu \neq 0$. □

2.5.4 Application 2: Global Bifurcation for Functional Differential Inclusions

We use the notion of phase space given in Sect. 1.4.

Set $I = [0, T]$ and denote by $\mathscr{BC}(\mathbf{R}^n)$ the Banach space of bounded continuous functions $BC((-\infty, 0]; \mathbf{R}^n)$.

Consider a functional differential inclusion with infinite delay of the following form:

$$x'(t) \in F(t, x_t, \mu) \ for \ a.e \ t \in [0, T], \qquad (2.46)$$

where $F: \mathbf{R} \times \mathscr{BC}(\mathbf{R}^n) \times \mathbf{R} \to Kv(\mathbf{R}^n)$ is a multimap.

Assume that the multimap F satisfies the next conditions:

($H1$) F is a T-periodic upper Carathéodory multimap.
($H2$) For every bounded subset $\Omega \subset C_T(I, \mathbf{R}^n) \times \mathbf{R}$ there exists a function $\nu_\Omega \in L_+^2[0, T]$ such that for each $(\varphi, \mu) \in \Omega$

$$\|F(t, \tilde{\varphi}_t, \mu)\|_{\mathbf{R}^n} \leq \nu_\Omega(t) \ for \ a.e. \ t \in [0, T],$$

where $\tilde{x}$ denotes the T-periodic extension of x on $(-\infty, T]$.
($H3$) $0 \in F(t, 0, \mu)$ for all $\mu \in \mathbf{R}$ and a.e. $t \in [0, T]$.

Notice that under conditions ($H1$)–($H2$) the superposition multioperator

$$\mathscr{P}_F: C_T(I, \mathbf{R}^n) \times \mathbf{R} \to Cv(L^2(I, \mathbf{R}^n)),$$

$$\mathscr{P}_F(x,\mu) = \{f \in L^2(I;\mathbf{R}^n): f(s) \in F(s, \tilde{x}_s, \mu) \text{ for a.e. } t \in I\},$$

is well-defined and closed.

As earlier, we can treat the global bifurcation problem of T-periodic solutions of inclusion (2.46) as the global bifurcation problem of solutions of the following operator inclusion

$$\ell x \in \mathscr{P}_F(x,\mu), \tag{2.47}$$

where ℓ is the operator of differentiation.

From $(H3)$ it follows that problem (2.47) has a trivial solution $(0, \mu)$ for all $\mu \in \mathbf{R}$. Let us denote by $\mathscr{S}$ the set of all nontrivial solutions of (2.47).

We use the notion of local integral guiding functions as given in Definition 2.20. Following the method given in the proof of Theorem 2.14 we obtain

Theorem 2.16. *Let conditions $(H1)$–$(H3)$ hold. Assume that for each μ with*

$$0 < |\mu - \mu_0| \le \varepsilon_0, \text{ where } \mu_0, \varepsilon_0 \text{ are given constants,}$$

there exists a local integral guiding function V_μ to problem (2.46) such that

$$\lim_{\mu \to \mu_0^+} ind\, V_\mu - \lim_{\mu \to \mu_0^-} ind\, V_\mu \neq 0.$$

Then there is a connected subset $\mathscr{R} \subset \mathscr{S}$ such that $(0, \mu_0) \in \overline{\mathscr{R}}$ and either $\mathscr{R}$ is unbounded or $\overline{\mathscr{R}} \ni (0, \mu_*)$ for some $\mu_* \neq \mu_0$.

2.5.5 Application 3: Feedback Control System

Consider the following feedback control system with infinite delay

$$\begin{cases} x'(t) = \mu a x(t) + f(x_t, u(t), \mu) \text{ for a.a. } t \in [0, T], \\ u(t) \in U(x(t)) \text{ for a.a. } t \in [0, T], \\ x(0) = x(T), \end{cases} \tag{2.48}$$

where $a > 0$, $\mu \in \mathbf{R}$, a map $f: \mathscr{BC}(\mathbf{R}^n) \times \mathbf{R}^m \times \mathbf{R} \to \mathbf{R}^n$ is continuous; a multimap $U: \mathbf{R}^n \to Kv(\mathbf{R}^m)$ is u.s.c.; $n, m \in \mathbf{N}$ and n is an odd number.

We assume the following conditions:

$(f1)$ There exist $\gamma > 1$ and $b > 0$ such that

$$|f(\tilde{\varphi}_t, y, \mu)| \le b \, \|\varphi\|_2^\gamma (|\mu| + |y|)$$

for all $(\varphi, y, \mu) \in C_T(I, \mathbf{R}^n) \times \mathbf{R}^m \times \mathbf{R}$ and a.e. $t \in [0, T]$.

$(U1)$ For every $(\varphi, \mu) \in \mathscr{BC}(\mathbf{R}^n) \times \mathbf{R}$ the set $f\left(\varphi, U(\varphi(0)), \mu\right)$ is convex.

(U2) There exists $c > 0$ such that

$$\|U(y)\|_{\mathbf{R}^m} \leq c(1 + |y|)$$

for all $y \in \mathbf{R}^n$.

Define a multimap $F: \mathscr{BC}(\mathbf{R}^n) \times \mathbf{R} \to Kv(\mathbf{R}^n)$ by

$$F(\varphi, \mu) = \mu a\varphi(0) + f(\varphi, U(\varphi(0)), \mu).$$

Then we treat the problem of global bifurcation of T-periodic solutions of problem (2.48) as the problem of global bifurcation of T-periodic solutions of the following differential inclusion:

$$x'(t) \in F(x_t, \mu), \quad \text{for a.e. } t \in I.$$

Let us denote by $\mathscr{S}$ the set of all nontrivial T-periodic solutions of (2.48).

Theorem 2.17. *Let conditions* $(f1)$ *and* $(U1) - (U2)$ *hold. Then there is a connected unbounded subset* $\mathscr{R} \subset \mathscr{S}$ *such that* $(0,0) \in \overline{\mathscr{R}}$.

Proof. It is easy to see that multimap F satisfies all conditions $(H1)$–$(H3)$ in Theorem 2.16. For each $\mu \neq 0$ consider the function

$$V_\mu: \mathbf{R}^n \to \mathbf{R},$$

$$V_\mu(y) = \frac{1}{2}\mu\langle y, y \rangle$$

Letting $x \in W_T^{1,2}(I, \mathbf{R}^n)$ and choosing an arbitrary $g \in \mathscr{P}_F(x, \mu)$, we obtain that there exists $u \in L^2(I, \mathbf{R}^m)$ such that $u(s) \in U(x(s))$ for a.e. $s \in I$, and

$$g(s) = \mu a x(s) + f(\tilde{x}_s, u(s), \mu) \text{ for a.e. } s \in I.$$

We have

$$\int_0^T \langle grad\, V_\mu(x(t)), g(t)\rangle dt = \int_0^T \langle \mu x(t), \mu a x(t) + f(\tilde{x}_t, u(t), \mu)\rangle dt$$

$$\geq a\mu^2\|x\|_2^2 - |\mu| \int_0^T |x(t)| |f(\tilde{x}_t, u(t), \mu)| dt$$

$$\geq a\mu^2\|x\|_2^2 - b|\mu|\|x\|_2^\gamma \int_0^T |x(t)|(|\mu| + |u(t)|) dt$$

$$\geq a\mu^2\|x\|_2^2 - b|\mu|\|x\|_2^\gamma \int_0^T |x(t)|(|\mu| + c + c|x(t)|) dt$$

$$\geq a\mu^2\|x\|_2^2 - b\sqrt{T}(\mu^2 + c|\mu|)\|x\|_2^{1+\gamma} - bc|\mu|\|x\|_2^{2+\gamma}.$$

Therefore

$$\int_0^T \langle grad\, V_\mu(x(t)), g(t)\rangle dt \geq |\mu| \|x\|_2^2 \big(a|\mu| - b\sqrt{T}(|\mu| + c)\|x\|_2^{\gamma-1} - bc\|x\|_2^\gamma\big) > 0,$$

$$(2.49)$$

for all $\mu \neq 0$ and sufficiently small $\|x\|_2 \neq 0$.

Thus for every $\mu \neq 0$, V_μ is a local integral guiding function for problem (2.48). From the fact that

$$\lim_{\mu \to 0^+} ind V_\mu - \lim_{\mu \to 0^-} ind V_\mu = 1 - (-1)^n = 2$$

and (2.49) it follows that $(0,0)$ is the unique bifurcation point for problem (2.48). Applying Theorem 2.16 we obtain the conclusion of the theorem.   □

Chapter 3
Method of Guiding Functions in Hilbert Spaces

In this chapter we present a new approach to extend the method of guiding function for differential and functional differential inclusions in Hilbert spaces. The results in this chapter were partly published in [100, 108, 109].

3.1 Integral Guiding Functions for Differential Inclusions in Hilbert Spaces

In this section we describe the extension of the method of integral guiding functions to the infinite dimensional case.

3.1.1 The Setting of the Problem

Let H be a Hilbert space with an orthonormal basis $\{e_n\}_{n=1}^{\infty}$. For every $n \geq 1$, let H_n be an n-dimensional subspace of H with the basis $\{e_k\}_{k=1}^{n}$ and P_n be a projection of H onto H_n. By $(\cdot, \cdot)_H$ we denote the inner product in H. The symbol I denotes the interval $[0, T]$. The symbol $(\cdot, \cdot)_{L^2}$ denote the inner product in $L^2(I, H)$.

The embedding $W^{1,2}(I, H) \hookrightarrow C(I, H)$ is continuous, and for every $n \geq 1$ the space $W^{1,2}(I, H_n)$ is compactly embedded in $C(I, H_n)$ (see, e.g., [14]). The weak convergence in $W^{1,2}(I, H)$ $[L^2(I, H)]$ is denoted by $x_n \overset{W}{\rightharpoonup} x_0$ [respectively, $f_n \overset{L^2}{\rightharpoonup} f_0$].

Let $n \geq 1$, and $\ell: W_T^{1,2}(I, H_n) \to L^2(I, H_n)$ be a linear Fredholm operator of index zero. Then there exist (see, Sect. 1.3) operators C_n, Q_n, Λ_n, Π_n and K_{C_n, Q_n} such that the equation

$$\ell x = y, \ y \in L^2(I, H_n)$$

is equivalent to

$$(i - C_n)x = (\Lambda_n \Pi_n + K_{C_n, Q_n})y.$$

The following notion plays an important role in the sequel.

Let $\mathscr{A}: W_T^{1,2}(I, H) \rightarrow L^2(I, H)$ be a linear operator; $\mathscr{F}: C_T(I, H) \rightarrow P(L^2(I, H))$ a multimap. For $n \geq 1$ define the projection $\mathbf{P}_n : L^2(I, H) \rightarrow L^2(I, H_n)$ generated by P_n as

$$(\mathbf{P}_n f)(t) = P_n f(t), \quad \text{for a.e. } t \in I.$$

Definition 3.1. An inclusion

$$\mathscr{A} x \in \mathscr{F}(x)$$

is said to be approximation solvable, if from the existence of sequences $\{n_k\}$ and $\{x^{(k)}\}$, $x^{(k)} \in W_T^{1,2}(I, H_{n_k})$ such that

$$\sup_k \|x^{(k)}\|_C < +\infty \quad \text{and} \quad \mathscr{A} x^{(k)} \in \mathbf{P}_{n_k} \mathscr{F}(x^{(k)})$$

it follows that there is a subsequence $\{x^{(k_m)}\}$ such that

$$x^{(k_m)} \xrightarrow{W} x^* \in W_T^{1,2}(I, H), \quad \text{and} \quad \mathscr{A} x^* \in \mathscr{F}(x^*).$$

Remark 3.1. Notice that the notion of an approximation solvable inclusion is closely related to the notion of A-proper map (see Proposition 21.3 in [38]). Properties and applications of A-proper maps were studied extensively in works [30, 110, 121, 131] and others.

Consider the differential inclusion

$$x'(t) \in F(t, x(t)), \text{ for a.e. } t \in I, \tag{3.1}$$

where $F: \mathbf{R} \times H \rightarrow Kv(H)$ is a multimap.

Suppose the following conditions are satisfied

(H1) F is T-periodic and upper Carathéodory.

(H2) For every $r > 0$ there is a function $v_r \in L_+^2[0, T]$ such that from $x \in C_T(I, H)$ and $\|x\|_2 \leq r$ it follows that

$$\|F(t, x(t))\|_H \leq v_r(t)$$

for a.a. $t \in I$.

By a T-periodic solution to problem (3.1) we mean a function $x \in W_T^{1,2}(I, H)$ such that there is a function $f \in L^2(I, H)$ satisfying conditions:

$$f(t) \in F(t, x(t)) \quad \text{and} \quad x'(t) = f(t) \text{ for a.a. } t \in I.$$

From conditions $(H1)$–$(H2)$ it follows that the superposition multioperator

$$\mathscr{P}_F : C_T(I, H) \to Cv(L^2(I, H)),$$

is well-defined and closed.

Then we can treat the existence of T-periodic solutions to problem (3.1) as the existence of solutions to the following operator inclusion

$$Ax \in \mathscr{P}_F(x), \qquad (3.2)$$

where $A : W_T(I, H) \to L^2(I, H)$ is the operator of differentiation.

Recall that a continuous differentiable function $V : H \to \mathbf{R}$ is said to be a non-degenerate potential, if there is $R_0 > 0$ such that

$$\nabla V(x) = \left(\frac{\partial V(x)}{\partial x_1}, \frac{\partial V(x)}{\partial x_2}, \cdots, \frac{\partial V(x)}{\partial x_n}, \cdots \right) \neq 0$$

for all $x = (x_1, x_2, \cdots, x_n, \cdots) \in H$ provided $\|x\|_H \geq R_0$.

Definition 3.2. A continuous differentiable function V is called projectively homogeneous potential if there exists $n_0 \in \mathbf{N}$ such that

$$Pr_n \nabla V(x) = \nabla V(P_n x)$$

for all $n \geq n_0$ and all $x \in H$, where $Pr_n : \mathbf{R}^\infty \to \mathbf{R}^\infty$ is the natural projection on first n coordinates.

Example 3.1. Set $V_1, V_2 : H \to \mathbf{R}$ be function defined as

$$V_1(x) = \langle x, x \rangle_H,$$

and

$$V_2(x) = \sum_{i=1}^{100} x_i x_{i+1} + \sum_{j=102}^{\infty} x_j^2, \ where \ x = (x_1, x_2, \cdots) \in H.$$

It is easy to verify that V_1 and V_2 are projectively homogeneous potentials.

For each $n \in \mathbf{N}$, let us make the natural identification $H_n \cong Pr_n \mathbf{R}^\infty \cong \mathbf{R}^n$. Then, restricting the multifield $Pr_n \nabla V$ on H_n, we can consider it as the continuous field $Pr_n \nabla V : \mathbf{R}^n \to \mathbf{R}^n$.

From Definition 3.2 it follows that if V is a non-degenerate projectively homogeneous potential then the fields $Pr_n \nabla V$ have no zeros on spheres $\partial B_{\mathbf{R}^n}(0, R)$ for all $n \geq n_0$ and $R \geq R_0$. So the topological degrees

$$\gamma_n = deg(Pr_n \nabla V, \partial B_{\mathbf{R}^n}(0, R)), \ n \geq n_0,$$

are well-defined and do not depend on $R \geq R_0$.

The index of the non-degenerate projectively homogeneous potential V is defined by:

$$Ind\ V = (\gamma_{n_0}, \gamma_{n_0+1}, \cdots).$$

By $Ind\ V \neq 0$ we mean that there exists a subsequence $\{n_k\}$ such that $\gamma_{n_k} \neq 0$ for all n_k.

Definition 3.3. A projectively homogenous potential $V: H \to \mathbf{R}$ is said to be an integral guiding function for inclusion (3.1) if there exists $N > 0$ such that for every $x \in W_T^{1,2}(I, H)$ from $\|x\|_2 \geq N$, it follows that

$$\overline{\lim}_{n \to \infty}\ sign \left(\int_0^T \langle Pr_n \nabla V(x(s)), f(s) \rangle\, ds \right) = 1$$

for all $f \in \mathscr{P}_F(x)$, where $\langle \cdot, \cdot \rangle$ denotes the inner product in $\mathbf{R}^n$.

Lemma 3.1. *If V is an integral guiding function for inclusion (3.1), then V is a non-degenerate potential.*

Proof. In fact, for every $y = (y_1, y_2, \cdots) \in H$, $\|y\|_H \geq \frac{N}{\sqrt{T}}$, considering y as a constant function we have that $\|y\|_2 \geq N$. Hence,

$$\overline{\lim}_{n \to \infty}\ sign \left(\int_0^T \langle Pr_n \nabla V(y), f(s) \rangle\, ds \right) = 1,$$

for all $f \in \mathscr{P}_F(y)$. So $\nabla V(y) \neq (0, 0, \cdots, 0, \cdots)$.   □

3.1.2 Existence of Periodic Solutions

We present a general application of the MGF to the existence of a periodic solution to a differential inclusion in a Hilbert space.

Theorem 3.1. *Let conditions $(H1) - (H2)$ hold. Assume that there exists an integral guiding function V for inclusion (3.1) such that $Ind\ V \neq 0$. If inclusion (3.2) is approximation solvable then inclusion (3.1) has a T-periodic solution.*

Remark 3.2. Some sufficient conditions of approximation solvability of inclusion (3.2) are given in Theorems 3.2 and 3.3.

Proof (of Theorem 3.1). It is easy to see that for each $n \geq 1$ the restriction

$$A_n = A|_{W_T^{1,2}(I, H_n)} : W_T^{1,2}(I, H_n) \to L^2(I, H_n)$$

is the linear Fredholm operator of index zero and

$$Ker\, A_n \cong H_n \cong Coker\, A_n.$$

The spaces $W_T^{1,2}(I, H_n)$ and $L^2(I, H_n)$ can be decomposed as:

$$W_T^{1,2}(I, H_n) = W_0^{(n)} \oplus W_1^{(n)},$$

and

$$L^2(I, H_n) = \mathscr{L}_0^{(n)} \oplus \mathscr{L}_1^{(n)},$$

where $W_0^{(n)} = ker\, A_n$, $\mathscr{L}_0^{(n)} = coker\, A_n$, $W_1^{(n)} = (W_0^{(n)})^{\perp}$ and $\mathscr{L}_1^{(n)} = Im A_n$. For every $u \in W_T^{1,2}(I, H_n)$ and $f \in L^2(I, H_n)$ we denote their corresponding decompositions by

$$u = u_{(0)}^{(n)} + u_{(1)}^{(n)},$$

and

$$f = f_{(0)}^{(n)} + f_{(1)}^{(n)}.$$

Notice that a function $x \in W_T^{1,2}(I, H_n)$ is a solution of the inclusion

$$A_n x \in \mathbf{P}_n \mathscr{P}_F(x)$$

if and only if it is a fixed point

$$x \in G_n(x), \tag{3.3}$$

of the multimap

$$G_n : C_T(I, H_n) \to Kv\big(C_T(I, H_n)\big),$$
$$G_n(x) = C_n x + (\Lambda_n \Pi_n + K_{C_n, Q_n}) \circ \mathbf{P}_n \mathscr{P}_F(x),$$

where projection $\Pi_n : L_2(I, H_n) \to H_n$ is defined as

$$\Pi_n f - \frac{1}{T} \int\limits_0^T f(s)\, ds$$

and the homomorphism $\Lambda_n : H_n \to H_n$ is the identity operator.

Following the method given in the proof of Theorem 2.14 it is easy to prove that the multimap G_n is completely upper semicontinuous. Now let us prove that the solutions of inclusion (3.2) are priori bounded in the space $C_T(I, H)$. In fact, assume that $\tilde{x} \in W_T^{1,2}(I, H)$ is a solution of inclusion (3.2). Then there is a function $\tilde{f} \in \mathscr{P}_F(\tilde{x})$ such that $\tilde{x}'(t) = \tilde{f}(t)$ for a.a. $t \in I$. Using the relation $Pr_n \nabla V(\tilde{x}(t)) = \nabla V(P_n \tilde{x}(t))$ for all $t \in I$ and $n \geq n_0$ (where n_0 is the coefficient occurring in Definition 3.2), we have

$$\overline{\lim}_{n \to \infty} sign\left(\int_0^T \langle Pr_n \nabla V(\tilde{x}(s)), \tilde{f}(s) \rangle \, ds \right)$$

$$= \overline{\lim}_{n \to \infty} sign\left(\int_0^T \langle \nabla V(P_n \tilde{x}(s)), P_n \tilde{x}'(s) \rangle \, ds \right)$$

$$= \overline{\lim}_{n \to \infty} sign\left(V(P_n \tilde{x}(T)) - V(P_n \tilde{x}(0)) \right) = 0,$$

Hence, $\|\tilde{x}\|_2 < N$, where N is the constant in Definition 3.3. From $(H2)$ it follows that there exists $K > 0$ such that $\|\tilde{x}'\|_2 < K$. Then there is a number $M > 0$, independent of $\tilde{x}$, such that $\|\tilde{x}\|_C < M$.

Choose an arbitrary $R \geq \max\{R_0, M\}$, where R_0 follows from the non-degeneracy of the potential V. Then inclusion (3.2) has no solutions on $\partial B_{C_T}(0, R)$. Let us show that for each $n \geq n_0$

$$x \notin G_n(x)$$

provided $x \in \partial B_{C_T}^{(n)}(0, R) = \partial B_{C_T}(0, R) \cap C_T(I, H_n)$.

To the contrary, assume that $x^* \in \partial B_{C_T}^{(n_*)}(0, R)$, $n_* \geq n_0$, is a solution of inclusion (3.3). Then there is a function $f^* \in \mathscr{P}_F(x^*)$ such that $Ax^* = \mathbf{P}_{n_*} f^*$. From the choice of R it follows that $\|x^*\|_2 \geq N$. Then we obtain

$$\overline{\lim}_{n \to \infty} sign\left(\int_0^T \langle Pr_n \nabla V(x_*(s)), f^*(s) \rangle \, ds \right) = 1.$$

Since the function x^* takes values in H_{n_*} and V is projectively homogeneous, we have

$$\overline{\lim}_{n \to \infty} sign\left(\int_0^T \langle Pr_n \nabla V(x_*(s)), f^*(s) \rangle \, ds \right)$$

$$= sign\left(\int_0^T \langle \nabla V(P_{n_*} x_*(s)), P_{n_*} f^*(s) \rangle \, ds \right)$$

$$= sign\left(\int_0^T \langle \nabla V(x_*(s)), x_*'(s) \rangle \, ds \right)$$

$$= sign\left(V(x_*(T)) - V(x_*(0)) \right) = 0,$$

that is the contradiction.

Thus, for each $n \geq n_0$ the topological degree

$$\omega_n = deg(i - G_n, B_{C_T}^{(n)}(0, R))$$

is well-defined.

Now we evaluate ω_n. For this we consider the multimap

$$\Sigma_n : C_T(I, H_n) \times [0, 1] \to Kv(C_T(I, H_n)),$$

$$\Sigma_n(x, \lambda) = C_n x + (\Lambda_n \Pi_n + K_{C_n, Q_n}) \circ \chi_n(\mathbf{P}_n \mathscr{P}_F(x), \lambda),$$

where $\chi_n : L^2(I, H_n) \times [0, 1] \to L^2(I, H_n)$ is defined as

$$\chi_n\left(f_{(0)}^{(n)} + f_{(1)}^{(n)}, \lambda\right) = f_{(0)}^{(n)} + \lambda f_{(1)}^{(n)}.$$

It is easy to see that the multimap Σ_n is completely upper semicontinuous. Let us show that the set

$$Fix\left(\Sigma_n, \partial B_{C_T}^{(n)}(0, R) \times [0, 1]\right)$$

of fixed points of the family $\Sigma_n(\cdot, \lambda)$ on $\partial B_{C_T}^{(n)}(0, R)$ is empty. To the contrary, assume that there exists $(x^*, \lambda^*) \in \partial B_{C_T}^{(n)}(0, R) \times [0, 1]$ such that

$$x^* \in \Sigma_n(x^*, \lambda^*).$$

Then there is a function $f^* \in \mathscr{P}_F(x^*)$ such that

$$\begin{cases} A_n x^* = \lambda^* f_{(1)}^{*(n)} \\ 0 = f_{(0)}^{*(n)}, \end{cases}$$

where $f_{(0)}^{*(n)} + f_{(1)}^{*(n)} = \mathbf{P}_n f^*$, $f_{(0)}^{*(n)} \in \mathscr{L}_0^{(n)}$ and $f_{(1)}^{*(n)} \in \mathscr{L}_1^{(n)}$.
It is clear that $\|x^*\|_2 \geq N$. Then we have

$$\overline{\lim}_{m \to \infty} sign\left(\int_0^T \langle Pr_m \nabla V(x^*(s)), f^*(s) \rangle ds\right) = 1.$$

Since $x^* \in C_T(I, H_n)$ we obtain

$$\overline{\lim}_{m \to \infty} sign\left(\int_0^T \langle Pr_m \nabla V(x^*(s)), f^*(s) \rangle ds\right)$$

$$= sign\left(\int_0^T \langle \nabla V(P_n x^*(s)), P_n f^*(s) \rangle ds\right)$$

$$= sign\left(\int_0^T \langle \nabla V(x^*(s)), P_n f^*(s) \rangle ds\right).$$

If $\lambda^* \neq 0$, then

$$sign\left(\int_0^T \langle \nabla V(x^*(s)), P_n f^*(s)\rangle ds\right) = sign\left(\int_0^T \langle \nabla V(x^*(s)), \frac{1}{\lambda^*} x^{*\prime}(s)\rangle ds\right)$$

$$= sign\left(\frac{1}{\lambda^*} \int_0^T \langle \nabla V(x^*(s)), x^{*\prime}(s)\rangle ds\right)$$

$$= sign\left(\frac{1}{\lambda^*}(V(x^*(T)) - V(x^*(0)))\right) = 0,$$

that is the contradiction.

In case $\lambda^* = 0$, we have $A_n x^* = 0$. Therefore, $x^* \in ker A_n$, i.e.,

$$x^*(t) \equiv y = (y_1, \cdots, y_n, 0, 0, \cdots), \quad t \in I,$$

where $\|y\|_H = R$.

From $\|y\|_2 \geq N$ it follows that

$$\overline{\lim}_{m \to \infty} sign\left(\int_0^T \langle Pr_m \nabla V(y), f(s)\rangle ds\right) = 1,$$

for all $f \in \mathscr{P}_F(y)$.

On the other hand

$$\overline{\lim}_{m \to \infty} sign\left(\int_0^T \langle Pr_m \nabla V(y), f(s)\rangle ds\right) = sign\left(\int_0^T \langle \nabla V(P_n y), P_n f(s)\rangle ds\right)$$

$$= sign\langle \nabla V(y), \int_0^T P_n f(s) ds\rangle$$

$$= sign\left(\langle \nabla V(y), \Pi_n f^{(n)}\rangle\right),$$

where $f^{(n)} = \mathbf{P}_n f \in \mathbf{P}_n \mathscr{P}_F(y)$. So

$$\langle \nabla V(y), \Pi_n f^{(n)}\rangle > 0, \tag{3.4}$$

and hence, $\Pi_n f^{(n)} \neq 0$ for all $f \in \mathscr{P}_F(y)$. In particular, $\Pi_n f^{*(n)} \neq 0$. But $\Pi_n f^{*(n)} = \Pi_n f_{(0)}^{*(n)} = 0$, giving the contradiction.

Thus, Σ_n is a homotopy connecting the multioperators:

$$\Sigma_n(x, 1) = G_n \text{ and } \Sigma_n(x, 0) = C_n + \Pi_n \mathbf{P}_n \mathscr{P}_F.$$

Then we obtain

$$deg\left(i - G_n, B_{C_T}^{(n)}(0, R)\right) = deg\left(i - C_n - \Pi_n \mathbf{P}_n \mathscr{P}_F, B_{C_T}^{(n)}(0, R)\right).$$

The operator $C_n + \Pi_n \mathbf{P}_n \mathscr{P}_F$ takes values in $H_n \cong \mathbf{R}^n$, so, by the map restriction property of the topological degree we obtain

$$deg\big(i - C_n - \Pi_n \mathbf{P}_n \mathscr{P}_F, B^{(n)}_{C_T}(0, R)\big) = deg\big(i - C_n - \Pi_n \mathbf{P}_n \mathscr{P}_F, B_{\mathbf{R}^n}(0, R)\big).$$

In the space $H_n \cong \mathbf{R}^n$ the multifield $i - C_n - \Pi_n \mathbf{P}_n \mathscr{P}_F$ has the form

$$i - C_n - \Pi_n \mathbf{P}_n \mathscr{P}_F = -\Pi_n \mathbf{P}_n \mathscr{P}_F,$$

therefore,

$$deg\big(i - C_n - \Pi_n \mathbf{P}_n \mathscr{P}_F, B_{\mathbf{R}^n}(0, R)\big) = deg\big(-\Pi_n \mathbf{P}_n \mathscr{P}_F, B_{\mathbf{R}^n}(0, R)\big).$$

From (3.4) it follows that the multifields $\Pi_n \mathbf{P}_n \mathscr{P}_F$ and $Pr_n \nabla V$ are homotopic on $B_{\mathbf{R}^n}(0, R)$, and then

$$deg\big(-\Pi_n \mathbf{P}_n \mathscr{P}_F, B_{\mathbf{R}^n}(0, R)\big) = deg\big(-Pr_n \nabla V, B_{\mathbf{R}^n}(0, R)\big) = (-1)^n \gamma_n.$$

From $Ind\ V \neq 0$ it follows that there exists a sequence $\{n_k\}$, $n_k \geq n_0$, such that $\gamma_{n_k} \neq 0$, and then $\omega_{n_k} \neq 0$. So, there is a sequence $\{x^{(k)}\}$, $x^{(k)} \in B^{(n_k)}_{C_T}(0, R)$, such that $Ax^{(k)} \in \mathbf{P}_{n_k} \mathscr{P}_F(x^{(k)})$ for all k. Using the approximation solvability of the inclusion (3.2) we obtain that inclusion (3.1) has a T-periodic solution. $\qquad \square$

3.1.3 Approximation Conditions

In this section we present some sufficient conditions for approximation solvability of inclusion (3.2).

Theorem 3.2. *Let a Hilbert space H be compactly embedded in a Banach space Y. Assume that the multimap $\tilde{F}: I \times Y \to P(Y)$ satisfies the following conditions:*

($\tilde{F}$1) for a.a. $t \in I$ the multimap $\tilde{F}(t, \cdot): Y \to P(Y)$ is u.s.c.;
($\tilde{F}$2) the restriction $F = \tilde{F}_{|I \times H}$ takes values in $Kv(H)$ and the multimap $F: I \times H \to Kv(H)$ is L^2-upper Carathéodory.

Then inclusion (3.2) is approximation solvable.

Proof. Assume that there are sequences $\{n_k\}$ and $\{x^{(k)}\}$, $x_k \in C_T(I, H_{n_k})$, such that

$$\sup_k \|x^{(k)}\|_C < +\infty \quad and \quad Ax^{(k)} \in \mathbf{P}_{n_k} \mathscr{P}_F(x^{(k)}).$$

From ($\tilde{F}$2) it follows that the set $\mathscr{P}_F(\{x^{(k)}\}_{k=1}^\infty)$, and hence the set $A(\{x^{(k)}\}_{k=1}^\infty)$, is bounded in $L^2(I, H)$. Then the set $\{x^{(k)}\}_{k=1}^\infty$ is bounded in $W_T^{1,2}(I, H)$, and so it is weakly compact.

W.l.o.g. assume that $x^{(k)} \xrightarrow{W} x^{(0)} \in W_T^{1,2}(I, H)$. Therefore, $Ax^{(k)} \xrightarrow{L^2} Ax^{(0)}$. From the fact that H is compactly embedded in Y it follows that the space $W_T^{1,2}(I, H)$ is compactly embedded in $C_T(I, Y)$, and hence,

$$x^{(k)} \xrightarrow{C_T(I,Y)} x^{(0)}. \tag{3.5}$$

Now let $f^{(k)} \in \mathscr{P}_F(x^{(k)})$ be such that $Ax^{(k)} = \mathbf{P}_{n_k} f^{(k)}$. The set $\{f^{(k)}\}_{k=1}^{\infty}$ is bounded in $L^2(I, H)$, so it is weakly compact in this space. W.l.o.g. assume that

$$f^{(k)} \xrightarrow{L^2} f^{(0)} \in L^2(I, H).$$

Let us show that $\mathbf{P}_{n_k} f^{(k)} \xrightarrow{L^2} f^{(0)}$. For this, at first we demonstrate that

$$\lim_{n \to \infty} \mathbf{P}_n f^{(0)} = f^{(0)}.$$

It fact, since

$$L^2(I, H) = \overline{\bigcup_{n=1}^{\infty} L^2(I, H_n)},$$

there are sequences $\{\hat{n}_m\}_{m=1}^{\infty}$ and $\{\hat{f}^{(m)}\}_{m=1}^{\infty}$, $\hat{f}^{(m)} \in L^2(I, H_{\hat{n}_m})$ such that $\hat{f}^{(m)} \to f^{(0)}$ in $L^2(I, H)$.
We have

$$\|\mathbf{P}_{\hat{n}_m} f^{(0)} - f^{(0)}\|_2 \le \|\mathbf{P}_{\hat{n}_m} f^{(0)} - \mathbf{P}_{\hat{n}_m} \hat{f}^{(m)}\|_2 + \|\mathbf{P}_{\hat{n}_m} \hat{f}^{(m)} - f^{(0)}\|_2$$

$$\le 2\|\hat{f}^{(m)} - f^{(0)}\|_2 \to 0$$

as $m \to \infty$. Further, for all $n > \hat{n}_m$

$$\|\mathbf{P}_n f^{(0)} - \mathbf{P}_{\hat{n}_m} f^{(0)}\|_2 = \|\mathbf{P}_n f^{(0)} - \mathbf{P}_n(\mathbf{P}_{\hat{n}_m} f^{(0)})\|_2 \le \|f^{(0)} - \mathbf{P}_{\hat{n}_m} f^{(0)}\|_2,$$

hence,

$$\|\mathbf{P}_n f^{(0)} - f^{(0)}\|_2 \le \|\mathbf{P}_n f^{(0)} - \mathbf{P}_{\hat{n}_m} f^{(0)}\|_2 + \|\mathbf{P}_{\hat{n}_m} f^{(0)} - f^{(0)}\|_2$$

$$\le 2\|f^{(0)} - \mathbf{P}_{\hat{n}_m} f^{(0)}\|_2.$$

So,

$$\lim_{n \to \infty} \mathbf{P}_n f^{(0)} = f^{(0)}.$$

Now for every $g \in L^2(I, H)$ we obtain

$$
\begin{aligned}
\left\langle \mathbf{P}_{n_k} f^{(k)} - f^{(0)}, g \right\rangle_{L^2} &= \left\langle \mathbf{P}_{n_k} f^{(k)} - \mathbf{P}_{n_k} f^{(0)}, g \right\rangle_{L^2} + \left\langle \mathbf{P}_{n_k} f^{(0)} - f^{(0)}, g \right\rangle_{L^2} \\
&= \left\langle f^{(k)} - f^{(0)}, \mathbf{P}_{n_k} g \right\rangle_{L^2} + \left\langle \mathbf{P}_{n_k} f^{(0)} - f^{(0)}, g \right\rangle_{L^2} \\
&= \left\langle f^{(k)} - f^{(0)}, g \right\rangle_{L_2} + \left\langle f^{(k)} - f^{(0)}, \mathbf{P}_{n_k} g - g \right\rangle_{L^2} \\
&\quad + \left\langle \mathbf{P}_{n_k} f^{(0)} - f^{(0)}, g \right\rangle_{L^2}.
\end{aligned}
$$

Thus

$$
\lim_{k \to \infty} \left\langle \mathbf{P}_{n_k} f_k - f_0, g \right\rangle_{L^2} = 0.
$$

On the other hand, $\mathbf{P}_{n_k} f^{(k)} = Ax^{(k)} \overset{L^2}{\rightharpoonup} Ax^{(0)}$. So $Ax^{(0)} = f^{(0)}$, and hence, $f^{(k)} \overset{L^2}{\rightharpoonup} Ax^{(0)}$. By the Mazur's Lemma (see, e.g., [44], p. 16) there is a sequence of convex combinations $\{\overline{f}^{(m)}\}$,

$$
\overline{f}^{(m)} = \sum_{k=m}^{\infty} \lambda_{mk} f^{(k)}, \; \lambda_{mk} \geq 0 \; and \; \sum_{k=m}^{\infty} \lambda_{mk} = 1,
$$

which converges to $Ax^{(0)}$ on average. Applying Theorem 38 ([126], Chap. IV), we assume w.l.o.g that $\{\overline{f}^{(m)}\}$ converges to Ax^0 for a.e. $t \in I$. Since the embedding $H \hookrightarrow Y$ is compact, we have $\overline{f}^{(m)}(t) \overset{Y}{\to} Ax^{(0)}(t)$ for a.e. $t \in I$.

From (3.5) and $(\tilde{F}1)$ it follows that for a.a. $t \in I$ and for a given $\varepsilon > 0$ there is an integer $i_0 = i_0(\varepsilon, t)$ such that

$$
\tilde{F}(t, x^{(i)}(t)) \subset O_{\varepsilon}^Y \left(\tilde{F}\left(t, x^{(0)}(t)\right) \right) \text{ for all } i \geq i_0,
$$

where O_{ε}^Y denotes the ε-neighborhood of a set in Y. Since $x^{(i)}(t) \in H$ for all i, we obtain

$$
F(t, x^{(i)}(t)) \subset O_{\varepsilon}^Y \left(F\left(t, x^{(0)}(t)\right) \right) \text{ for all } i \geq i_0,
$$

Then $f^{(i)}(t) \in O_{\varepsilon}^Y \left(F\left(t, x^{(0)}(t)\right) \right)$ for all $i \geq i_0$, and by the convexity of the set $O_{\varepsilon}^Y \left(F\left(t, x^{(0)}(t)\right) \right)$ we have

$$
\overline{f}^{(m)}(t) \in O_{\varepsilon}^Y \left(F\left(t, x^{(0)}(t)\right) \right), \text{ for all } m \geq i_0.
$$

Therefore, $Ax^{(0)}(t) \in F(t, x^{(0)}(t))$ for a.e. $t \in I$, and so

$$
Ax^{(0)} \in \mathscr{P}_F(x^{(0)}).
$$

$\square$

Theorem 3.3. *Let $F: I \times H \to Kv(H)$ be a L^2-upper Carathéodory multimap. Then inclusion (3.2) is approximation solvable in each of the following cases:*

(1i) for a.e. $t \in I$ the multimap $F(t, \cdot): H \to Kv(H)$ is weakly u.s.c. in the following sense: for every sequence $\{\psi^{(n)}\} \in H$, $\psi^{(n)} \overset{H}{\rightharpoonup} \psi^{(0)} \in H$, and for every $\varepsilon > 0$ there is an integer $N(\varepsilon, t) > 0$ such that

$$F(t, \psi^{(n)}) \subset O_\varepsilon\big(F(t, \psi^{(0)})\big)$$

for all $n > N(\varepsilon, t)$;

(2i) there is an integer $q_0 > 0$ such that for each $n \geq q_0$ the restriction of $F(t, \cdot)$ on H_n takes values in $Kv(H_n)$ for a.a. $t \in I$.

Proof. Assume that there are sequences $\{n_k\}_{k=1}^{+\infty}$ and $\{x^{(k)}\}$, $x^{(k)} \in C_T(I, H_{n_k})$, such that

$$\sup_k \|x^{(k)}\|_C < +\infty \quad and \quad Ax^{(k)} \in \mathbf{P}_{n_k}\mathscr{P}_F(x^{(k)}).$$

Let condition (1i) holds. Then the multioperator $\mathscr{P}_F$ is well-defined. Similarly to the proof of Theorem 3.2, from $x^{(k)} \overset{W}{\rightharpoonup} x^{(0)}$ it follows that $x^{(k)}(t) \overset{H}{\rightharpoonup} x^{(0)}(t)$, for almost $t \in I$. And hence, from condition (1i) we obtain that for a.e. $t \in I$

$$F(t, x^{(i)}(t)) \subset O_\varepsilon\big(F\big(t, x^{(0)}(t)\big)\big) \text{ for all } i \geq N(t, \varepsilon).$$

Hence once again we have $Ax^{(0)} \in \mathscr{P}_F(x^{(0)})$.

Now let condition (2i) holds. Then for each $n \geq q_0$ we obtain

$$\mathbf{P}_n \mathscr{P}_F(x) = \mathscr{P}_F(x),$$

for all $x \in C_T(I, H_n)$. It is clear that for all k such that $n_k \geq q_0$ the following relation is satisfied: $Ax^{(k)} \in \mathscr{P}_F(x^{(k)})$. $\square$

3.1.4 Application 1: Control Problem of a Partial Differential Equation

In this section we consider the control problem of the following partial differential equation

$$\begin{cases} \frac{\partial u(t,s)}{\partial t} = b + au(t, s) + \int_0^1 K(t, s, \sigma)\big[u(t, \sigma) + v(\sigma)\big]d\sigma, \\ v \in U, \end{cases} \qquad (3.6)$$

for all $s \in [0, 1]$ and a.a. $t \in \mathbf{R}$ where $a > 0$, $b \in \mathbf{R}$, $K: \mathbf{R} \times [0, 1] \times [0, 1] \to \mathbf{R}$ is a continuous map and $U \subset L^1[0, 1]$ is a bounded closed subset.

By a 1-periodic solution to problem (3.6) we mean a continuous function $u: \mathbf{R} \times [0, 1] \to \mathbf{R}$ such that:

(a) u is T-periodic with respect to the first argument.
(b) The partial derivative $\frac{\partial u(t,s)}{\partial t}$ is a Carathéodory function.
(c) There exists a function $v \in U$ such that the pair (u, v) satisfies (3.6).

Let us denote by $Y = C[0, 1]$ and $H = W^{1,2}[0, 1]$. It is clear that Y is a separable Banach space, H is a Hilbert space with an inner product

$$\langle u, v \rangle_H = \int_0^1 u(s)v(s)ds + \int_0^1 u'(s)v'(s)ds,$$

the embedding $H \hookrightarrow Y$ is compact and $\|z\|_Y \leq \|z\|_H$ for all $z \in H$. Let $\{e_1, e_2, \cdots\}$ be a orthonormal basis in H. We will consider problem (3.6) with the following hypothesis.

(K1) K is 1-periodic with respect to the first argument.
(K2) $\frac{\partial K(t,s,\sigma)}{\partial s}$ is a continuous function.

For each $t \in [0, 1]$ let us denote by $x(t) = u(t, \cdot)$. From the continuity of u it follows the continuity of $x: [0, 1] \to Y$. Moreover, if $x'(t)$ exists for a.e. $t \in [0, 1]$ and the function x' is measurable, then there exists $\frac{\partial u(t,s)}{\partial t}$ which is Carathéodory. Therefore, problem (3.6) can be substituted by the following problem

$$\begin{cases} x'(t) \in b + ax(t) + G(t, x(t)), & \text{for a.a. } t \in [0, 1], \\ x(0) = x(1), \end{cases} \tag{3.7}$$

where $G: \mathbf{R} \times Y \to Kv(H)$,

$$G(t, y) = \left\{ f \in H: f(s) = \int_0^1 K(t, s, \sigma)[y(\sigma) + v(\sigma)]d\sigma; \ v \in U \right\}.$$

Theorem 3.4. *Let conditions* (K1) – (K2) *hold. Then inclusion (3.6) has a T-periodic solution for each sufficiently large a.*

Proof. Define $\tilde{F}: \mathbf{R} \times Y \to Kv(Y)$ by

$$\tilde{F}(t, y) = b + ay + G(t, y).$$

It is easy to verify that $\tilde{F}$ satisfies all condition in Theorem 3.2 and the multimap $F - \tilde{F}|_{\mathbf{R} \times H}$ satisfies conditions (H1) – (H2). Set

$$V: H \to \mathbf{R},$$

$$V(z) = \sum_{n=1}^{\infty} \frac{1}{2} c_n^2,$$

where c_n are coordinates of z with respect to basis $\{e_1, e_2, \cdots\}$. It is clear that V is a projectively homogeneous potential. Let us show that V is an integral guiding function for problem (3.7).

In fact, letting $x \in W_T^{1,2}(I, H)$ and choosing arbitrarily $v \in U$, then

$$\tilde{f}(t) = b + ax(t) + g(t) \in \tilde{F}(t, x(t)), \text{ for a.e. } t \in [0, 1],$$

where

$$g(t)(s) = \int_0^1 K(t, s, \sigma)[x(t)(\sigma) + v(\sigma)]d\sigma, \text{ for } t \in [0, 1]. \tag{3.8}$$

Notice that for each $t \in [0, 1]$: $w = x(t)$ and $z = g(t)$ are functions in H. We have

$$\langle x(t), b + ax(t) + g(t) \rangle_H = \langle w, b + aw + z \rangle_H$$

$$= \int_0^1 a\big(w^2(\tau) + w'^2(\tau)\big)d\tau + b \int_0^1 w(\tau)d\tau$$

$$+ \int_0^1 \big(w(\tau)z(\tau) + w'(\tau)z'(\tau)\big)d\tau$$

$$\geq a\|w\|_H^2 - |b|\|w\|_H - \|w\|_H\|z\|_H.$$

Therefore,

$$\int_0^1 \langle \nabla V(x(t)), \tilde{f}(t) \rangle_H ds = \int_0^1 \langle x(t), b + ax(t) + g(t) \rangle_H ds$$

$$\geq \int_0^1 \big(a\|x(t)\|_H^2 - |b|\|x(t)\|_H - \|g(t)\|_H\|x(t)\|_H\big) ds.$$

Since (3.8) and the boundedness of the set U there exist $M, L > 0$ such that

$$\|g(t)\|_H \leq L + M\|x(t)\|_H, \text{ for } t \in [0, 1].$$

Consequently,

$$\int_0^1 \langle \nabla V(x(t)), \tilde{f}(t) \rangle_H ds \geq (a - M)\|x\|_2^2 - (|b| + L)\|x\|_2 > 0$$

provided $\|x\|_2 > \frac{|b|+L}{a-M}$. So,

$$\overline{\lim}_{n \to \infty} sign \left(\int_0^1 \langle P_n \nabla V(x(t)), \tilde{f}(t) \rangle_H ds \right)$$

$$= sign \left(\int_0^1 \langle \nabla V(x(t)), \tilde{f}(t) \rangle_H ds \right) = 1.$$

From Theorem 3.1 it follows that inclusion (3.6) has a 1-periodic solution for each $a > M$. □

3.2 Non-smooth Guiding Functions for Functional Differential Inclusions with Infinite Delay in Hilbert Spaces

In the present section, developing the approach given above (see Sections 2.2 and 2.3), we define the notion of the non-smooth integral guiding function for a system governed by a functional differential inclusion with infinite delay in a Hilbert space and study the existence of periodic oscillations in such systems.

3.2.1 Setting of the Problem

Let H be a Hilbert space. We use the notion of phase space given in Sect. 1.4. The Banach space $BC((-\infty, 0]; H)$ of bounded continuous functions is denoted by $\mathscr{BC}(H)$.

We suppose to study the functional differential inclusion in H with the infinite delay of the following form

$$x'(t) \in F(t, x_t) \text{ for a.a. } t \in I. \tag{3.9}$$

We assume that a multimap $F: \mathbf{R} \times \mathscr{BC}(H) \to Kv(H)$ satisfies the following conditions:

$(H1)$ F is T-periodic upper Carathéodory.
$(H2)$ for every $r > 0$ there exists a function $v_r \in L^2_+[0, T]$ such that for each $x \in C_T(I, H)$ with $\|x\|_2 \leq r$ we have

$$\|F(s, \tilde{x}_s)\|_H \leq v_r(s) \text{ for a.a. } s \in I,$$

where $\tilde{x}$ denotes the T-periodic extension of x on $(-\infty, T]$.

From above conditions it follows that the superposition multioperator

$$\mathscr{P}_F: C_T(I, H) \to Cv(L^2(I, H)),$$

$$\mathscr{P}_F(x) = \{f \in L^2(I, H): f(s) \in F(s, \tilde{x}_s) \text{ for a.a. } s \in I\}$$

is well-defined and closed.

Then we treat the problem of existence of T-periodic solutions of inclusion (3.9) as the problem of existence of solutions of the following operator inclusion

$$Ax \in \mathscr{P}_F(x), \tag{3.10}$$

where A is the operator of differentiation.

In the sequel, we use some notions of non-smooth analysis given in Chap. 1.

Given a regular function $V: H \to \mathbf{R}$, for each $i = 1, 2, \ldots$, define the function

$$V_i: \mathbf{R} \to \mathbf{R}, \ V_i(y) = V(0, \cdots, 0, y, 0, \cdots),$$

where y is placed in the i-th position. It is clear that V_i is also regular.

We define *the generalized gradient* $\partial^* V(x)$ of a regular function V at the point $x = (x_1, x_2, \cdots) \in H$ in the following way:

$$\partial^* V(x) = \partial V_1(x_1) \times \partial V_2(x_2) \times \ldots \times \partial V_i(x_i) \times \ldots \subset \mathbf{R}^\infty,$$

where $\partial V_i, i = 1, 2, \ldots$ is the subdifferential of the function V_i.

Let us note that our definition of *the generalized gradient* is different from the Clarke's generalized gradient and its calculation is easier. For example, let $V: \ell_2 \to \mathbf{R}$ be defined as

$$V(x) = |x_1| + |x_1 x_2 \cdots x_{100}| + \sum_2^\infty x_k^2, \ x = (x_1, x_2, \cdots). \tag{3.11}$$

We have that $\partial^* V(x) = \partial V_1(x_1) \times \{2x_2\} \times \cdots \times \{2x_n\} \times \cdots$, where

$$\partial V_1(x_1) = \begin{cases} \{1\}, & \text{if } x_1 > 0, \\ \{[-1, 1]\}, & \text{if } x_1 = 0, \\ \{-1\}, & \text{if } x_1 < 0. \end{cases}$$

However, the Clarke's generalized gradient is not so easy to calculate.

Definition 3.4. A regular function $V: H \to \mathbf{R}$ is said to be a projectively homogeneous potential, if there exists $n_0 \in \mathbf{N}$ such that

$$Pr_n \partial^* V(x) = \partial^* V(P_n x)$$

for all $n \geq n_0$ and $x \in H$.

It is easy to see that the function in (3.11) is projectively homogeneous.

Definition 3.5. A regular function $V: H \to \mathbf{R}$ is said to be a non-degenerate potential, if there exists $R_0 > 0$ such that

$$(0, 0, \cdots, 0, \cdots) \notin \partial^* V(x)$$

for all $x \in H$ such that $\|x\|_H \geq R_0$.

From Definitions 3.4 and 3.5 it follows that if V is a non-degenerate projectively homogeneous potential then the multifields $Pr_n \partial^* V$ have no singular points on spheres $\partial B_{\mathbf{R}^n}(0, R)$ for all $n \geq n_0$ and $R \geq R_0$. So the topological degrees

$$\gamma_n = deg(Pr_n \partial^* V, \partial B_{\mathbf{R}^n}(0, R)), \ n \geq n_0,$$

are well-defined and do not depend on $R \geq R_0$.

The index of the non-degenerate projectively homogeneous potential V is defined by:

$$Ind \ V = (\gamma_{n_0}, \gamma_{n_0+1}, \cdots).$$

By *Ind $V \neq 0$* we mean that there exists a subsequence $\{n_k\}$ such that $\gamma_{n_k} \neq 0$ for all n_k.

For every continuous function $x \in C(I, H)$, $x(t) = (x_1(t), x_2(t), \cdots)$, $t \in I$, by a selection $v(t) \in \partial^* V(x(t))$ we mean

$$v(t) = (v_1(t), v_2(t), \cdots), \ t \in I,$$

where $v_i(t) \in \partial V_i(x_i(t))$, for a.e. $t \in I$, $i \geq 1$, are summable selections.

Definition 3.6. A projectively homogeneous potential $V: H \to \mathbf{R}$ is said to be a non-smooth guiding function for inclusion (3.9), if there exists $N > 0$ such that for every $x \in W_T^{1,2}(I, H)$ from $\|x\|_2 \geq N$ it follows that:

$$\overline{\lim}_{m \to \infty} \ sign \left(\sum_{k=1}^m \int_0^T v_k(s) f_k(s) \, ds \right) = 1,$$

for all $f \in \mathscr{P}_F(x)$, $f(s) = (f_1(s), f_2(s), \ldots)$ and all selections $v(s) \in \partial^* V(x(s))$.

Lemma 3.2. *If V is a non-smooth guiding function for inclusion (3.9) then V is the non-degenerate potential.*

Proof. In fact, for every $y = (y_1, y_2, \cdots) \in H$, $\|y\|_H \geq \frac{N}{\sqrt{T}}$, considering y as the constant function we have that $\|y\|_2 \geq N$. Hence,

$$\overline{\lim}_{m \to \infty} \ sign \left(\sum_{k=1}^m \int_0^T v_k f_k(s) \, ds \right) = 1,$$

for all $f \in \mathscr{P}_F(y)$ and all $v = (v_1, v_2, \cdots) \in \partial^* V(y)$. So $v \neq (0, 0, \cdots, 0, \cdots)$. $\square$

3.2.2   Existence Theorem

In this section we present a result on the existence of a periodic solution for a functional differential inclusion in a Hilbert space.

Theorem 3.5. *Let conditions $(H1) - (H2)$ hold. Assume that there exists a non-smooth guiding function V for the inclusion (3.9) such that $\mathrm{Ind}\, V \neq 0$. If the inclusion (3.10) is approximation solvable then inclusion (3.9) admits a T-periodic solution.*

Proof. As in the previous section, for every $n \geq 1$ a function $x \in W_T^{1,2}(I, H_n)$ is a solution of the inclusion

$$A_n x \in \mathbf{P}_n \mathscr{P}_F(x)$$

if and only if it is a fixed point

$$x \in G_n(x), \tag{3.12}$$

of the following completely u.s.c. multimap

$$G_n : C_T(I, H_n) \to Kv\big(C_T(I, H_n)\big),$$

$$G_n(x) = C_n x + (\Lambda_n \Pi_n + K_{C_n, Q_n}) \circ \mathbf{P}_n \mathscr{P}_F(x).$$

Now let us demonstrate that solutions of inclusion (3.10) are priori bounded in the space $C_T(I, H)$. In fact, assume that $x \in W_T^{1,2}(I, H)$ is a solution of inclusion (3.10). Then there is a function $f \in \mathscr{P}_F(x)$ such that $x'(t) = f(t)$ for a.e. $t \in I$. For every selection $\upsilon(s) \in \partial^* V(x(s))$ we have

$$\overline{\lim}_{m \to \infty} \, sign \left(\sum_{k=1}^m \int_0^T \upsilon_k(s) f_k(s) \, ds \right)$$

$$= \overline{\lim}_{m \to \infty} \, sign \left(\sum_{k=1}^m \int_0^T \upsilon_k(s) x_k'(s) \, ds \right)$$

$$\leq \overline{\lim}_{m \to \infty} \, sign \left(\sum_{k=1}^m \int_0^T V_k^0\big(x_k(s), x_k'(s)\big) \, ds \right)$$

$$= \overline{\lim}_{m \to \infty} \, sign \left(\sum_{k=1}^m (V_k(x_k(T)) - V_k(x_k(0))) \right) = 0,$$

where $x(t) = (x_1(t), x_2(t), \cdots)$ and $f(t) = (f_1(t), f_2(t), \cdots), t \in I$.

Hence, $\|x\|_2 < N$. From $(H2)$ it follows that there exists $K > 0$ such that $\|x'\|_2 < K$. Then there is a number $M > 0$, independent of x, such that $\|x\|_C < M$. Choose an arbitrary $R \geq \max\{R_0, M\}$, where R_0 is the constant in Definition 3.5. Then inclusion (3.10) has no solutions on $\partial B_C(0, R)$. Let us show that for each $n \geq n_0$

$$x \notin G_n(x)$$

provided $x \in \partial B_C^{(n)}(0, R) = \partial B_C(0, R) \cap C_T(I, H_n)$.

To the contrary, assume that $x^* \in \partial B_C^{(n_*)}(0, R)$, $n_* \geq n_0$, is a solution of inclusion (3.12). Then there is a function $f^* \in \mathscr{P}_F(x^*)$ such that $Ax^* = \mathbf{P}_{n_*} f^*$. From the choice of R it follows that $\|x^*\|_2 \geq N$. Then we obtain

$$\overline{\lim}_{m \to \infty} sign \left(\sum_{k=1}^{m} \int_0^T \upsilon_k(s) f_k^*(s) \, ds \right) = 1,$$

for all selections $\upsilon(s) \in \partial^* V(x^*(s))$, $s \in I$.

Since the function x^* takes values in H_{n_*} and V is projectively homogeneous, we have

$$\overline{\lim}_{m \to \infty} sign \left(\sum_{k=1}^{m} \int_0^T \upsilon_k(s) f_k^*(s) \, ds \right) = sign \left(\sum_{k=1}^{n_*} \int_0^T \upsilon_k(s) f_k^*(s) \, ds \right)$$

$$= sign \left(\sum_{k=1}^{n_*} \int_0^T \upsilon_k(s) x_k^{*\prime}(s) \, ds \right)$$

$$\leq sign \left(\sum_{k=1}^{n_*} \int_0^T V_k^0 \left(x_k^*(s), x_k^{*\prime}(s) \right) \, ds \right)$$

$$= sign \left(\sum_{k=1}^{n_*} \left(V_k(x_k^*(T)) - V_k(x_k^*(0)) \right) \right) = 0,$$

that is a contradiction.

Thus, for each $n \geq n_0$ the topological degree

$$\omega_n = deg(i - G_n, B_C^{(n)}(0, R))$$

is well-defined.

Now we evaluate ω_n. For this purpose, we consider the multimap

$$\Sigma_n : C_T(I, H_n) \times [0, 1] \to Kv(C_T(I, H_n)),$$

$$\Sigma_n(x, \lambda) = C_n x + (\Lambda_n \Pi_n + K_{C_n, Q_n}) \circ \chi_n(\mathbf{P}_n \mathscr{P}_F(x), \lambda),$$

where χ_n is defined as in the previous section.

It is easy to see that the multimap Σ_n is completely u.s.c.. Let us show that the set

$$Fix\left(\Sigma_n, \partial B_C^{(n)}(0, R) \times [0, 1] \right)$$

of fixed points of the family $\Sigma_n(\cdot, \lambda)$ on $\partial B_C^{(n)}(0, R)$ is empty. To the contrary, assume that there exists $(x^*, \lambda^*) \in \partial B_C^{(n)}(0, R) \times [0, 1]$ such that

$$x^* \in \Sigma_n(x^*, \lambda^*).$$

Then there is a function $f^* \in \mathscr{P}_F(x^*)$ such that

$$\begin{cases} A_n x^* = \lambda^* f_{(1)}^{*(n)} \\ 0 = f_{(0)}^{*(n)}, \end{cases}$$

where $f_{(0)}^{*(n)} + f_{(1)}^{*(n)} = \mathbf{P}_n f^*$, $f_{(0)}^{*(n)} \in \mathscr{L}_0^{(n)}$ and $f_{(1)}^{*(n)} \in \mathscr{L}_1^{(n)}$.
It is clear that $\|x^*\|_2 \geq N$. Then we have

$$\overline{\lim}_{m \to \infty} \, sign \left(\sum_{k=1}^{m} \int_0^T \upsilon_k(s) f_k^*(s) \, ds \right) = 1,$$

for all selections $\upsilon(s) \in \partial^* V(x^*(s))$, $s \in I$.
Since $x^* \in C_T(I, H_n)$ we obtain

$$\overline{\lim}_{m \to \infty} \, sign \left(\sum_{k=1}^{m} \int_0^T \upsilon_k(s) f_k^*(s) \, ds \right) = sign \left(\sum_{k=1}^{n} \int_0^T \upsilon_k(s) f_k^*(s) \, ds \right),$$

where $f^*(t) = (f_1^*(t), f_2^*(t), \cdots)$ and $x^*(t) = (x_1^*(t), \cdots, x_n^*(t), 0, 0, \cdots)$.
If $\lambda^* \neq 0$, then

$$sign \left(\sum_{k=1}^{n} \int_0^T \upsilon_k(s) f_k^*(s) \, ds \right) = sign \left(\frac{1}{\lambda^*} \sum_{k=1}^{n} \int_0^T \upsilon_k(s) x_k^{*\prime}(s) \, ds \right)$$

$$\leq sign \left(\sum_{k=1}^{n} \int_0^T V_k^0(x_k^*(s), x_k^{*\prime}(s)) \, ds \right)$$

$$= sign \left(\sum_{k=1}^{n} (V_k(x_k^*(T)) - V_k(x_k^*(0))) \right) = 0,$$

that is a contradiction.
In case $\lambda^* = 0$, we have $A_n x^* = 0$. Therefore, $x^* \in ker A_n$, i.e.,

$$x^*(t) \equiv y = (y_1, \cdots, y_n, 0, 0, \cdots), \quad t \in I,$$

where $\|y\|_H = R$.

From the fact that From $\|y\|_2 \geq N$ it follows that

$$\overline{\lim}_{m \to \infty} \, sign \left(\sum_{k=1}^{m} \int_{0}^{T} \upsilon_k \, f_k(s) \, ds \right) = 1,$$

for all $f \in \mathscr{P}_F(y)$ and all elements $\upsilon = (\upsilon_1, \cdots, \upsilon_n, 0, 0, \cdots) \in \partial^* V(y)$.
On the other hand

$$\overline{\lim}_{m \to \infty} \, sign \left(\sum_{k=1}^{m} \int_{0}^{T} \upsilon_k \, f_k(s) \, ds \right) = sign \left(\sum_{k=1}^{n} \int_{0}^{T} \upsilon_k \, f_k(s) \, ds \right)$$

$$= sign \left\langle \upsilon, \int_{0}^{T} (\mathbf{P}_n f)(s) ds \right\rangle = sign \langle \upsilon, \Pi_n f^{(n)} \rangle,$$

where $f^{(n)} = \mathbf{P}_n f \in \mathbf{P}_n \mathscr{P}_F(y)$. So

$$\langle \upsilon, \Pi_n f^{(n)} \rangle > 0, \tag{3.13}$$

and hence, $\Pi_n f^{(n)} \neq 0$ for all $f \in \mathscr{P}_F(y)$. In particular, $\Pi_n f^{*(n)} \neq 0$. But $\Pi_n f^{*(n)} = \Pi_n f_{(0)}^{*(n)} = 0$, giving the contradiction.

Thus, Σ_n is a homotopy connecting the multioperators $\Sigma_n(x, 1) = G_n$ and $\Sigma_n(x, 0) = C_n + \Pi_n \mathbf{P}_n \mathscr{P}_F$. Analogously Theorem 3.1 we obtain that

$$deg\left(i - G_n, B_C^{(n)}(0, R)\right) = deg\left(-Pr_n \partial^* V, B_{\mathbf{R}^n}(0, R)\right) = (-1)^n \gamma_n.$$

Then we obtain that inclusion (3.9) has a T-periodic solution.   □

Generalizing the results given in the previous section, Theorems 3.2 and 3.3, let us present some sufficient conditions for approximation solvability of inclusion (3.10).

For a Banach space Y, let us denote by $\mathscr{BC}(Y)$ the Banach space of all bounded continuous functions $x: (-\infty, 0] \to Y$.

Theorem 3.6. *Let a Hilbert space H be compactly embedded in a Banach space Y. Assume that the multimap $\tilde{F}: I \times \mathscr{BC}(Y) \to P(Y)$ satisfies the following conditions:*

($\tilde{F}$1) *for a.e. $t \in I$ the multimap $\tilde{F}(t, \cdot): \mathscr{BC}(Y) \to P(Y)$ is upper semicontinuous;*

($\tilde{F}$2) *the restriction $F = \tilde{F}_{|I \times \mathscr{BC}(H)}$ takes values in $Kv(H)$ and multimap $F: I \times \mathscr{BC}(H) \to Kv(H)$ is upper Carathéodory satisfied condition $(H2)$.*

Then inclusion (3.10) is approximation solvable.

Theorem 3.7. *Let $F: I \times \mathscr{BC}(H) \to Kv(H)$ be an upper Carathéodory multimap satisfying condition $(H2)$. Then inclusion (3.10) is approximation solvable in each of the following cases:*

(1i) *for a.e. $t \in I$ the multimap $F(t,\cdot): \mathscr{BC}(H) \to Kv(H)$ is weakly upper semicontinuous in the following sense: for every sequence $\{\psi^{(n)}\} \in H$, $\psi^{(n)} \overset{\mathscr{BC}(H)}{\rightharpoonup} \psi^{(0)} \in \mathscr{BC}(H)$, and for every $\varepsilon > 0$ there is an integer $N(\varepsilon,t) > 0$ such that*

$$F(t,\psi^{(n)}) \subset O_\varepsilon\big(F(t,\psi^{(0)})\big)$$

for all $n > N(\varepsilon,t)$;

(2i) *there is an integer $q_0 > 0$ such that for each $n \geq q_0$ the restriction of $F(t,\cdot)$ on $\mathscr{BC}(H_n)$ takes values in $Kv(H_n)$ for a.a. $t \in I$.*

3.2.3   Application: Existence of Periodic Solutions for a Gradient Functional Differential Inclusion

For $h > 0$, consider the spaces of real-valued functions $H = W^{1,2}[0,h]$ and $Y = L^2[0,h]$. It is clear that H is compactly embedded in Y. Let the functional $V: Y \to \mathbf{R}$ be defined as

$$V(y) = \frac{1}{2}|y_1| + \sum_{1}^{\infty} y_k^2, \quad y = (y_1, y_2, \cdots),$$

where $y_i, i = 1, 2, \cdots$, are the Fourier's coefficients of y. It is clear that

$$\partial^* V(y) = \partial V_1(y_1) \times \{2y_2\} \times \{2y_3\} \times \cdots,$$

where

$$\partial V_1(y_1) = \begin{cases} \{2y_1 + \frac{1}{2}\} : y_1 > 0, \\ [-\frac{1}{2}, \frac{1}{2}] : \quad y_1 = 0, \\ 2y_1 - \frac{1}{2} : \quad y_1 < 0, \end{cases}$$

and the multimap $\partial^* V: Y \to Kv(Y)$ is upper semicontinuous. Moreover, the restriction $\partial^* V_{|H}$ takes values in $Kv(H)$ and

$$\|\partial^* V(y)\|_H \leq 2\|y\|_H + \frac{1}{2}, \text{ for all } y = (y_1, y_2, \cdots) \in H. \tag{3.14}$$

Consider the following functional differential inclusion

$$x'(t) \in \partial^* V(x(t)) + G(t, x_t), \text{ for a.e. } t \in I, \tag{3.15}$$

where $G: \mathbf{R} \times \mathscr{BC}(Y) \to P(Y)$ is a T-periodic multimap.
Assume that the following conditions hold:

(G1) For a.e. $t \in I$ multimap $G(t, \cdot): \mathscr{BC}(Y) \to P(Y)$ is upper semicontinuous.
(G2) The restriction $G_{|_{I \times \mathscr{BC}(H)}}$ takes values in $Kv(H)$.
(G3) For each $\psi \in \mathscr{BC}(H)$ the multifunction $G(\cdot, \psi): I \to Kv(H)$ has a measurable selection.
(G4) There exists $C > 0$ such that

$$\|G(s, \tilde{\psi}_s)\|_H \leq C(1 + \|\psi\|_2),$$

for a.e. $s \in I$ and all $\psi \in C_T(I, H)$.

Theorem 3.8. *Let conditions* (G1) − (G4) *hold. In addition, assume that*

$$C \sqrt{T} < 2.$$

Then inclusion (3.15) has a T-periodic solution $x \in W_T^{1,2}(I, H)$.

Proof. Set $\tilde{F}: \mathbf{R} \times \mathscr{BC}(Y) \to P(Y)$,

$$\tilde{F}(t, \psi) = \partial^* V(\psi(0)) + G(t, \psi).$$

It is clear that the multimap $\tilde{F}$ is T-periodic with respect to the first argument and satisfies condition $(\tilde{F}1)$ of Theorem 3.6.
Set $F = \tilde{F}_{|_{I \times \mathscr{BC}(H)}}$. It is easy to see that the multimap F takes values in $Kv(H)$ and the multimap $F: I \times \mathscr{BC}(H) \to Kv(H)$ is upper Carathéodory. The role of condition $(H2)$ is that for every $r > 0$ and $x \in C_T(I, H)$ such that $\|x\|_2 \leq r$, there exists $M_r > 0$ such that $\|f\|_2 \leq M_r$ for all $f \in \mathscr{P}_F(x)$. From (3.14) and $(G4)$ we see that the multimap F satisfies condition $(H2)$. The application of Theorem 3.6 implies that the inclusion (3.10) is approximation solvable.

It is clear that the functional V is projectively homogeneous. Let us show that it is a guiding function for inclusion (3.15). In fact, let $x \in W_T^{1,2}(I, H)$ and take an arbitrary $f \in \mathscr{P}_F(x)$. Then there are a function $g \in \mathscr{P}_G(x)$ and a selection $\upsilon(s) \in \partial^* V(x(s))$ such that

$$f(s) = \upsilon(s) + g(s) \text{ for a.e. } t \in I.$$

Notice that for every $s \in I$ the values $u = \upsilon(s)$ and $\omega = g(s)$ are functions in H and

$$\langle v(s), f(s) \rangle_H = \langle u, u + \omega \rangle_H$$

$$= \int_0^h \left(u^2(\tau) + {u'}^2(\tau) \right) d\tau + \int_0^h \left(u(\tau)\omega(\tau) + u'(\tau)\omega'(\tau) \right) d\tau$$

$$\geq \|u\|_H^2 - \|u\|_H \|\omega\|_H.$$

Therefore

$$\int_0^T \langle v(s), f(s) \rangle_H ds = \int_0^T \langle v(s), v(s) + g(s) \rangle_H ds$$

$$\geq \int_0^T \left(\|v(s)\|_H^2 - \|g(s)\|_H \|v(s)\|_H \right) ds$$

$$\geq \|v\|_2^2 - \int_0^T \|v(s)\|_H \, C(1 + \|x\|_2) ds.$$

From (3.14) it follows that

$$\int_0^T \langle v(s), f(s) \rangle_H ds \geq \|v\|_2^2 - C(1 + \|x\|_2) \int_0^T \left(2\|x(s)\|_H + \frac{1}{2} \right) ds$$

$$\geq \|v\|_2^2 - 2C\sqrt{T}\|x\|_2^2 - \left(2C\sqrt{T} + \frac{TC}{2} \right)\|x\|_2 - \frac{TC}{2}.$$

Now let us mention that for every selection $v(s) \in \partial^* V(x(s))$ there is a number $\varepsilon \in [-\frac{1}{2}, \frac{1}{2}]$ such that

$$v(s) = (2x_1(s) + \varepsilon, 2x_2(s), \cdots, 2x_n(s), \cdots), \quad s \in I,$$

where $x(s) = (x_1(s), x_2(s), \cdots, x_n(s), \cdots), s \in I$.
Therefore

$$\|v\|_2^2 = \int_0^T \|v(s)\|_H^2 ds = 4 \int_0^T \|x(s)\|_H^2 ds + 4\varepsilon \int_0^T x_1(s) ds + \varepsilon^2 T$$

$$> 4\|x\|_2^2 - 2 \int_0^T \|x(s)\|_H ds \geq 4\|x\|_2^2 - 2\sqrt{T}\|x\|_2.$$

Hence we obtain

$$\int_0^T \langle v(s), f(s) \rangle_H ds > (4 - 2C\sqrt{T})\|x\|_2^2 - \left(2\sqrt{T} + 2C\sqrt{T} + \frac{TC}{2} \right)\|x\|_2 - \frac{TC}{2} > 0$$

provided $\|x\|_2$ is sufficiently large. So

$$\overline{\lim}_{m \to \infty} \, sign \left(\int_0^T \sum_{k=1}^m v_k(s) f_k(s) ds \right) = 1.$$

Thus, V is a guiding function for inclusion (3.15). It is clear that $Ind \, V \neq 0$. So, applying Theorem 3.5, we conclude that inclusion (3.15) has a T-periodic solution $x \in W_T^{1,2}(I, H)$. $\qquad\qquad\square$

3.3 Bifurcation Problem

In this section, applying the method of guiding functions, developed above, we study the global bifurcation problems for periodic solutions of parameterized ordinary differential inclusions in Hilbert spaces. As application we consider the global structure of a set of periodic trajectories for a family of feedback control systems.

3.3.1 The Setting of the Problem

Consider the following family of differential inclusions

$$\begin{cases} x'(t) \in F(t, x(t), \mu) \text{ for a.a. } t \in I, \\ x(0) = x(T), \end{cases} \tag{3.16}$$

where $F: \mathbf{R} \times H \times \mathbf{R} \to Kv(H)$ be a multimap.

We assume the following conditions:

$(H1)$ F is T-periodic L^2-upper Carathéodory;
$(H2)$ $0 \in F(t, 0, \mu)$ for all $\mu \in \mathbf{R}$ and a.a. $t \in I$.

We can substitute problem (3.16) by the following operator inclusion

$$Ax \in \mathscr{P}_F(x, \mu), \tag{3.17}$$

where $\mathscr{P}_F: C_T(I, H) \times \mathbf{R} \to Cv(L^2(I, H))$ is the superposition multioperator and $A: W_T^{1,2}(I, H) \to L^2(I, H)$ is the operator of differentiation.

From $(H2)$ it follows that $(0, \mu)$ is a solution to (3.17) for all $\mu \in \mathbf{R}$. Let us denote by $\mathscr{S}$ the set of all nontrivial solutions (x, μ), $x \neq 0$, to problem (3.17).

Definition 3.7 (cf. Definition 3.1). The inclusion (3.17) is said to be approximation solvable, if from the existence of sequences $\{n_k\}$ and $\{(x^{(k)}, \mu_k)\}$, $x^{(k)} \in W_T^{1,2}(I, H_{n_k})$, $\mu_k \in \mathbf{R}$ such that

$$\sup_k \|x^{(k)}\|_C < +\infty, \ \sup_k |\mu_k| < +\infty \ and$$

$$Ax^{(k)} \in \mathbf{P}_{n_k} \mathscr{P}_F(x^{(k)}, \mu_k),$$

it follows that there is a subsequence $\{(x^{(k_m)}, \mu_{k_m})\}$ such that

$$x^{(k_m)} \xrightarrow{W} x^* \in W_T^{1,2}(I, H), \ \mu_{k_m} \to \mu^* \in \mathbf{R}$$

and

$$Ax^* \in \mathscr{P}_F(x^*, \mu^*).$$

Definition 3.8. A continuously differentiable function $V: H \to \mathbf{R}$ is said to be a local non-degenerate potential, if there exists $R_0 > 0$ such that

$$\nabla V(x) \neq (0, 0, \cdots, 0, \cdots)$$

for all $x \in H$ such that $0 < \|x\|_H \leq R_0$.

Definition 3.9. A projectively homogenous potential $V_\mu: H \to \mathbf{R}$, depending on μ, is said to be a local integral guiding function for inclusion (3.16) at $(0, \mu_0)$, if there exists $\varepsilon_0 > 0$ such that for every μ, $0 < |\mu - \mu_0| < \varepsilon_0$ there is $\delta_\mu = \delta(\mu) > 0$, continuously depending on μ, such that from $x \in W_T^{1,2}(I, H)$ with one of the following conditions:

(i) $0 < \|x\|_2 < \delta_\mu$;
(ii) $0 < \|x\|_C < \delta_\mu$;

it follows that:

$$\overline{\lim}_{m \to \infty} \ sign \left(\int_0^T \left\langle Pr_m \nabla V_\mu(x(s)), f(s) \right\rangle ds \right) = 1,$$

for all $f \in \mathscr{P}_F(x, \mu)$.

Remark 3.3. If V is a local integral guiding function for inclusion (3.16) then V is the local non-degenerate potential.

Here we study the global structure of $\mathscr{S}$ near the given point $(0, \mu_0)$ in the case when inclusion (3.16) has a *local integral guiding function with condition (i)* [case (i)] and when inclusion (3.16) has a *local integral guiding function with condition (ii)* [case (ii)].

To do this for every $0 < \varepsilon < \varepsilon_0$ and sufficiently small $r > 0$ with

$$r < \min\{\delta_{\mu_0 - \varepsilon}, \delta_{\mu_0 + \varepsilon}, \frac{\delta_{\mu_0 - \varepsilon}}{\sqrt{T}}, \frac{\delta_{\mu_0 + \varepsilon}}{\sqrt{T}}\},$$

we define the vector fields

$$V_n^\sharp \colon \overline{U}_{r,\varepsilon}^{(n)} \to \mathbf{R}^n \times \mathbf{R}, \; n \geq n_0,$$

$$V_n^\sharp(y, \mu) = \{-Pr_n\nabla V_\mu(y), \varepsilon^2 - (\mu - \mu_0)^2\} = \{-\nabla V_\mu(y), \varepsilon^2 - (\mu - \mu_0)^2\},$$

where

$$\overline{U}_{r,\varepsilon}^{(n)} = \{(y, \mu) \in \mathbf{R}^n \times \mathbf{R} \colon |y|^2 + (\mu - \mu_0)^2 \leq r^2 + \varepsilon^2\}.$$

It is clear that $V_n^\sharp$ are continuous compact vector fields. Let us show that the fields $V_n^\sharp$ have no zeros on $\partial U_{r,\varepsilon}^{(n)}$, for all $n \geq n_0$.

Indeed, assume to the contrary that there is $(y, \mu) \in \partial U_{r,\varepsilon}^{(n)}$ such that $V_n^\sharp(y, \mu) = 0$. Then we obtain

$$\begin{cases} \mu = \mu_0 \pm \varepsilon \\ \nabla V_\mu(y) = 0. \end{cases}$$

Considering y as a constant function in $W_T^{1,2}(I, H_n)$, from the fact that V_μ is projectively homogenous potential and the choice of r it follows that $\|y\|_2 < \delta_\mu$ for case (i) (or $\|y\|_C < \delta_\mu$ for case (ii)). Hence for every $f \in \mathscr{P}_F(y, \mu)$ we have

$$1 = \overline{\lim}_{m\to\infty} sign\left(\int_0^T \left\langle Pr_m\nabla V_\mu(y), f(s)\right\rangle ds\right)$$

$$= sign\left\langle \nabla V_\mu(y), \int_0^T f(s)ds\right\rangle = 0,$$

giving a contradiction.

Therefore, the topological degrees

$$v_n = deg(V_n^\sharp, \overline{U}_{r,\varepsilon}^{(n)}), \; n \geq n_0,$$

are well defined. It is easy to see that for a given $\varepsilon \in (0, \varepsilon_0)$ these topological degrees do not depend on the choice of r and for given numbers $\varepsilon_1, \varepsilon_2 \in (0, \varepsilon_0)$ there exists $r_0 > 0$ such that

$$deg(V_n^\sharp, \overline{U}_{r,\varepsilon_1}^{(n)}) = deg(V_n^\sharp, \overline{U}_{r,\varepsilon_2}^{(n)})$$

for all $r \in (0, r_0)$ and all $n \geq n_0$.

Definition 3.10. The collection $(v_{n_0}, v_{n_0+1}, \cdots)$ is called the index of the map

$$V^\sharp \colon H \times \mathbf{R} \to \mathbf{R} \times \mathbf{R},$$

$$V^\sharp(y, \mu) = \{-V_\mu(y), \varepsilon^2 - (\mu - \mu_0)^2\},$$

at $(0, \mu_0)$ and is denoted by $ind\, V^\sharp (0, \mu_0)$.

By $ind\, V^\sharp (0, \mu_0) \neq 0$ we mean that there exists a subsequence $\{n_k\}$ such that $v_{n_k} \neq 0$ for all n_k.

3.3.2 Global Bifurcation Theorem

The following theorem on the structure of branches of non-trivial periodic solutions to a family of inclusions is the main result of this section.

Theorem 3.9. *Let conditions $(H1) - (H2)$ hold. In addition, assume that:*

$(H3)$ *inclusion (3.17) is approximation solvable;*
$(H4)$ *there exists a local integral guiding function V_μ to problem (3.16) at $(0, \mu_0)$*
       such that $ind\, V^\sharp (0, \mu_0) \neq 0$.

Then there is a connected subset $\mathscr{R} \subset \mathscr{S}$ such that $(0, \mu_0) \in \overline{\mathscr{R}}$ and at least one of the following occurs:

(a) $\mathscr{R}$ is unbounded;
(b) $(0, \mu_*) \in \overline{\mathscr{R}}$ for some $\mu_* \neq \mu_0$.

Remark 3.4. Some sufficient conditions for approximation solvability of inclusion (3.17) can be found in Theorems 3.2 and 3.3.

Proof. Consider the inclusion

$$A_n x \in \mathbf{P}_n \mathscr{P}_F(x, \mu)$$

or equivalently,

$$x \in G_n(x, \mu), \tag{3.18}$$

where

$$G_n \colon C_T(I, H_n) \times \mathbf{R} \to Kv(C_T(I, H_n)),$$

$$G_n(x, \mu) = C_n x + (\Lambda_n \Pi_n + K_{C_n, Q_n}) \circ \mathbf{P}_n \mathscr{P}_F(x, \mu).$$

STEP 1. It is easy to see that the multimaps G_n are completely u.s.c.
For $n \geq n_0$ and $r, \varepsilon > 0$ set

$$G_n^r \colon B_{r,\varepsilon}^{(n)} \to Kv(C_T(I, H_n) \times \mathbf{R}),$$

$$G_n^r(x, \mu) = \{x - G_n(x, \mu), \|x\|_C^2 - r^2\},$$

where

$$B_{r,\varepsilon}^{(n)} = \left\{(x, \mu) \in C_T(I, H_n) \times \mathbf{R} \colon \|x\|_C^2 + (\mu - \mu_0)^2 \leq r^2 + \varepsilon^2\right\}.$$

G_n^r are completely u.s.c. vector multifields. Choosing arbitrarily $\varepsilon \in (0, \varepsilon_0)$ and sufficiently small $r > 0$ with $r < \min\{\delta_{\mu_0 - \varepsilon}, \delta_{\mu_0 + \varepsilon}, \frac{\delta_{\mu_0 - \varepsilon}}{\sqrt{T}}, \frac{\delta_{\mu_0 + \varepsilon}}{\sqrt{T}}\}$, where ε_0 is the constant from Definition 3.9, we prove that $0 \notin G_n^r(x, \mu)$ for all $(x, \mu) \in \partial B_{r,\varepsilon}^{(n)}$.

Indeed, assume to the contrary that there is $(x, \mu) \in \partial B_{r,\varepsilon}^{(n)}$ such that $0 \in G_n^r(x, \mu)$. Then,

$$x \in G_n(x, \mu) \tag{3.19}$$

and

$$\|x\|_C = r. \tag{3.20}$$

From (3.19) it follows that there is $f \in \mathscr{P}_F(x, \mu)$ such that for a.a. $t \in I$:

$$x'(t) = P_n f(t).$$

From (3.20) and the fact that $(x, \mu) \in \partial B_{r,\varepsilon}^{(n)}$ we obtain that $\mu = \mu_0 \pm \varepsilon_0$. Moreover, $0 < \|x\|_2 \le \sqrt{T} \|x\|_C < \delta_\mu$ for case (i) $\left(\text{or } \|x\|_C = r < \delta_\mu \text{ for case } (ii) \right)$. Hence,

$$\overline{\lim}_{m \to \infty} \, sign \left(\int_0^T \left\langle Pr_m \nabla V_\mu(x(s)), f(s) \right\rangle ds \right) = 1.$$

Since $n \ge n_0$ and $x \in C_T(I, H_n)$ we have

$$1 = \overline{\lim}_{m \to \infty} \, sign \left(\int_0^T \left\langle Pr_m \nabla V_\mu(x(s)), f(s) \right\rangle ds \right)$$

$$= \overline{\lim}_{m \to \infty} \, sign \left(\int_0^T \left\langle \nabla V_\mu(P_m x(s)), f(s) \right\rangle ds \right)$$

$$= sign \left(\int_0^T \left\langle \nabla V_\mu(P_n x(s)), f(s) \right\rangle ds \right)$$

$$= sign \left(\int_0^T \left\langle \nabla V_\mu(x(s)), P_n f(s) \right\rangle ds \right)$$

$$= sign \left(\int_0^T \left\langle \nabla V_\mu(x(s)), x'(s) \right\rangle ds \right),$$

$$= V_\mu(x(T)) - V_\mu(x(0)) = 0$$

that is a contradiction.

Thus, for every $n \ge n_0$ the topological degree

$$\omega_n = deg(G_n^r, B_{r,\varepsilon}^{(n)})$$

is well-defined.

STEP 3. Now, we evaluate ω_n for $n \geq n_0$. Toward this goal, consider the multimap

$$\Sigma_n \colon B_{r,\varepsilon}^{(n)} \times [0,1] \to Kv(C_T(I, H_n) \times \mathbf{R}),$$

$$\Sigma_n(x, \mu, \lambda) = \left\{ x - C_n x - (\Lambda_n \Pi_n + K_{C_n, Q_n}) \circ \chi_n\big(\mathbf{P}_n \mathscr{P}_F(x, \mu), \lambda\big), \tau \right\},$$

$$\tau = \lambda(\|x\|_C^2 - r^2) + (1 - \lambda)(\varepsilon^2 - (\mu - \mu_0)^2),$$

where χ_n is defined as in the previous section.

It is easy to see that Σ_n is completely u.s.c. multifield. Let us show that

$$0 \notin \Sigma_n(x, \mu, \lambda)$$

for all $(x, \mu, \lambda) \in \partial B_{r,\varepsilon}^{(n)} \times [0,1]$.

To the contrary, let us assume that there exists $(\tilde{x}, \tilde{\mu}, \tilde{\lambda}) \in \partial B_{r,\varepsilon}^{(n)} \times [0,1]$ such that

$$0 \in \Sigma_n(\tilde{x}, \tilde{\mu}, \tilde{\lambda}).$$

Then

$$\tilde{\lambda}(\|\tilde{x}\|_C^2 - r^2) + (1 - \tilde{\lambda})(\varepsilon^2 - (\tilde{\mu} - \mu_0)^2) = 0 \qquad (3.21)$$

and

$$\begin{cases} A_n \tilde{x} = \tilde{\lambda}\, \tilde{f}_1^{(n)} \\ 0 = \tilde{f}_0^{(n)}, \end{cases}$$

where $\tilde{f}_0^{(n)} + \tilde{f}_1^{(n)} = \mathbf{P}_n \tilde{f}$ for some $\tilde{f} \in \mathscr{P}_F(\tilde{x}, \tilde{\mu})$, $\tilde{f}_0^{(n)} \in \mathscr{L}_0^{(n)}$ $\tilde{f}_1^{(n)} \in \mathscr{L}_1^{(n)}$. From $(\tilde{x}, \tilde{\mu}) \in \partial B_{r,\varepsilon}^{(n)}$ it follows that

$$\|\tilde{x}\|_C^2 - r^2 = \varepsilon^2 - (\tilde{\mu} - \mu_0)^2.$$

Hence, from (3.21) we obtain

$$\|\tilde{x}\|_C = r \ and \ \tilde{\mu} = \mu_0 \pm \varepsilon.$$

If $\tilde{\lambda} > 0$, then from the choice of r and the fact that $\tilde{x} \in C_T(I, H_n)$ we have

$$1 = \overline{\lim}_{m \to \infty} sign\left(\int_0^T \Big\langle Pr_m \nabla V_{\tilde{\mu}}(\tilde{x}(s)), \tilde{f}(s) \Big\rangle ds\right)$$

$$= \overline{\lim}_{m \to \infty} sign\left(\int_0^T \Big\langle \nabla V_{\tilde{\mu}}(P_m \tilde{x}(s)), \tilde{f}(s) \Big\rangle ds\right)$$

$$= sign\left(\int_0^T \Big\langle \nabla V_{\tilde{\mu}}(\tilde{x}(s)), P_n \tilde{f}(s) \Big\rangle ds\right)$$

$$= sign\left(\int_0^T \left\langle \nabla V_{\tilde{\mu}}(\tilde{x}(s)), \frac{1}{\lambda}\tilde{x}'(s)\right\rangle ds\right)$$

$$= sign\left(\int_0^T \left\langle \nabla V_{\tilde{\mu}}(\tilde{x}(s)), \tilde{x}'(s)\right\rangle ds\right)$$

$$= sign\left(V_{\tilde{\mu}}(\tilde{x}(T)) - V_{\tilde{\mu}}(\tilde{x}(0))\right) = 0,$$

giving the contradiction.

If $\tilde{\lambda} = 0$, then $A_n\tilde{x} = 0$, i.e., $\tilde{x}(t) = y \in H_n \cong \mathbf{R}^n$ for all $t \in I$. Since $\|y\|_2 < \delta_{\tilde{\mu}}$ for case (i) (or $\|y\|_C = r < \delta_\mu$ for case (ii)) we have

$$\overline{\lim}_{m\to\infty} sign\left(\int_0^T \left\langle Pr_m \nabla V_{\tilde{\mu}}(y), f(s)\right\rangle ds\right) = 1,$$

for all $f \in \mathscr{P}_F(y, \tilde{\mu})$.
On the other hand

$$\overline{\lim}_{m\to\infty} sign\left(\int_0^T \left\langle Pr_m \nabla V_{\tilde{\mu}}(y), f(s)\right\rangle ds\right) = sign\left(\int_0^T \left\langle \nabla V_{\tilde{\mu}}(P_n y), P_n f(s)\right\rangle ds\right)$$

$$= sign\left\langle \nabla V_{\tilde{\mu}}(y), \int_0^T P_n f(s)ds\right\rangle$$

$$= sign\left(\left\langle \nabla V_{\tilde{\mu}}(y), \Pi_n f^{(n)}\right\rangle\right),$$

where $f^{(n)} = P_n f \in P_n\mathscr{P}_F(y, \tilde{\mu})$. Therefore

$$\left\langle \nabla V_{\tilde{\mu}}(y), \Pi_n f^{(n)}\right\rangle > 0, \tag{3.22}$$

for all $f \in \mathscr{P}_F(y, \tilde{\mu})$.
In particular, $\Pi_n \tilde{f}^{(n)} \neq 0$, where $\tilde{f}^{(n)} = P_n \tilde{f}$. But $\Pi_n \tilde{f}^{(n)} = \Pi_n \tilde{f}_0^{(n)} = 0$, that is the contradiction.

Thus, Σ_n is a homotopy connecting the multifields $\Sigma_n(x, \mu, 1) = G_n^r(x, \mu)$ and

$$\Sigma_n(x, \mu, 0) = \{x - C_n x - \Pi_n P_n \mathscr{P}_f(x, \mu), \varepsilon^2 - (\mu - \mu_0)^2\}.$$

The homotopy invariance property of the topological degree implies that

$$deg\left(G_n^r, B_{r,\varepsilon}^{(n)}\right) = deg\left(\Sigma_n(\cdot, \cdot, 0), B_{r,\varepsilon}^{(n)}\right).$$

The operator $C_n + \Pi_n P_n \mathscr{P}_F$ takes values in $H_n \cong \mathbf{R}^n$, so

$$deg\left(\Sigma_n(\cdot, \cdot, 0), B_{r,\varepsilon}^{(n)}\right) = deg\left(\Sigma_n(\cdot, \cdot, 0), \overline{U}_{r,\varepsilon}^{(n)}\right),$$

where $\overline{U}_{r,\varepsilon}^{(n)} = B_{r,\varepsilon}^{(n)} \cap (\mathbf{R}^n \times \mathbf{R})$.

In the space $H_n \times \mathbf{R} \cong \mathbf{R}^n \times \mathbf{R}$ the vector field $\Sigma(\cdot, \cdot, 0)$ has the form

$$\Sigma_n(x, \mu, 0) = \{-\Pi_n \mathbf{P}_n \mathscr{P}_F(x, \mu), \varepsilon^2 - (\mu - \mu_0)^2\},$$

Consider now the multimap: $\Gamma : \overline{U}_{r,\varepsilon}^{(n)} \times [0, 1] \to Kv(\mathbf{R}^n \times \mathbf{R})$ defined by

$$\Gamma(y, \mu, \lambda) = \{-\lambda \Pi_n \mathbf{P}_n \mathscr{P}_F(y, \mu) + (\lambda - 1)\nabla V_\mu(y), \varepsilon^2 - (\mu - \mu_0)^2\}.$$

It is clear that Γ is a compact u.s.c. multifield. Assume that there exists $(x, \mu, \lambda) \in \partial U_{r,\varepsilon}^{(n)} \times [0, 1]$ such that $0 \in \Gamma(x, \mu, \lambda)$. Then we obtain

$$\begin{cases} \mu = \mu_0 \pm \varepsilon \\ (\lambda - 1)\nabla V_\mu(y) \in \lambda \Pi_n \mathbf{P}_n \mathscr{P}_F(y, \mu), \end{cases}$$

and by the (3.22) we get the contradiction. So, Γ is a homotopy connecting $\Sigma(\cdot, \cdot, 0)$ and $V_n^\sharp$, therefore

$$\omega_n = deg\big(\Sigma_n(\cdot, \cdot, 0), \overline{U}_{r,\varepsilon}^{(n)}\big) = deg\big(V_n^\sharp, \overline{U}_{r,\varepsilon}^{(n)}\big) = v_n. \qquad (3.23)$$

STEP 4. In this step following the method given in [113] we consider the global structure of solutions to problem (3.16).

For sufficiently small $r, \varepsilon > 0$, from (3.23) and $ind\ V^\sharp \neq 0$ it follows that there is a subsequence $\{n_k\}$ such that $\omega_{n_k} = v_{n_k} \neq 0$. Therefore, there exists the corresponding sequence $\{(x_{n_k}, \mu_{n_k})\} \subset B_{r,\varepsilon}^{(n_k)}$ such that $0 \in G_{n_k}^r(x_{n_k}, \mu_{n_k})$, or equivalently,

$$\begin{cases} x_{n_k} \in G_{n_k}(x_{n_k}, \mu_{n_k}) \\ \|x_{n_k}\|_C = r. \end{cases}$$

Condition $(H3)$ implies that there is (x_*, μ_*), $\|x_*\|_C = r$, such that $Ax_* \in \mathscr{P}_F(x_*, \mu_*)$. So, $(0, \mu_0)$ is a bifurcation point.

Let $\mathscr{O} \subset C_T(I, H) \times \mathbf{R}$ be an open subset defined as

$$\mathscr{O} = \big(C_T(I, H) \times \mathbf{R}\big) \setminus \big(\{0\} \times (\mathbf{R} \setminus (\mu_0 - \varepsilon_0, \mu_0 + \varepsilon_0))\big).$$

Let us denote by $\mathscr{W} \subset \big(\mathscr{S} \cup \{(0, \mu_0)\}\big) \subset \mathscr{O}$ the component of $(0, \mu_0)$. Assume that $\mathscr{W}$ is compact. Then there exists an open bounded subset $U \subset \mathscr{O}$ such that

$$\overline{U} \subset \mathscr{O}, \ \mathscr{W} \subset U \text{ and } \partial U \cap \mathscr{S} = \emptyset.$$

Hence, w.l.o.g. we can assume that $0 \notin G_n^r(x, \mu)$ for every $r > 0$ provided $(x, \mu) \in \partial U^{(n)}$, $\forall n \geq n_0$, where

$$\partial U^{(n)} = \partial U \cap (C_T(I, H_n) \times \mathbf{R}).$$

Further, for every $n \geq n_0$ and $R, r > 0$ consider the compact multifield $G_n^{\lambda r + (1-\lambda)R}$ on $\overline{U}^{(n)} \times [0, 1]$.

Assume that there exist sequences $\{n_k\}$ and $\{(x_{n_k}, \mu_{n_k}, \lambda_{n_k})\}$,

$$(x_{n_k}, \mu_{n_k}, \lambda_{n_k}) \in \partial U^{(n_k)} \times [0, 1],$$

such that

$$0 \in G_{n_k}^{\lambda_{n_k} r + (1 - \lambda_{n_k})R}(x_{n_k}, \mu_{n_k}, \lambda_{n_k}).$$

Then

$$\begin{cases} x'_{n_k} = \mathbf{P}_{n_k} f, \ f \in \mathscr{P}_F(x_{n_k}, \mu_{n_k}), \\ \|x_{n_k}\|_C = \lambda_{n_k} r + (1 - \lambda_{n_k})R. \end{cases}$$

From the approximation solvability property of inclusion (3.17) it follows that we can assume w.l.o.g. that

$$x_{n_k} \overset{W}{\to} x_0, \ \mu_{n_k} \to \mu_0, \ \lambda_{n_k} \to \lambda_0 \colon (x_0, \mu_0, \lambda_0) \in \partial U \times [0, 1]$$

with $Ax_0 \in \mathscr{P}_F(x_0, \mu_0)$, giving the contradiction.

So w.l.o.g., we can assume that

$$0 \notin G_n^{\lambda r + (1-\lambda)R}(x, \mu, \lambda)$$

for all $n \geq n_0$ provided $(x, \mu, \lambda) \in \partial U^{(n)} \times [0, 1]$.

Therefore, for every $n \geq n_0$ the multifields G_n^r and G_n^R are homotopic on $\partial U^{(n)}$. For sufficiently large R, the multifield G_n^R has no zeros on $\overline{U}^{(n)}$, so

$$deg(G_n^R, \overline{U}^{(n)}) = 0.$$

Thus, $deg(G_n^r, \overline{U}^{(n)}) = 0$ for every $r > 0$.

Let $\Lambda = \{\mu \in \mathbf{R} \colon (0, \mu) \in \overline{U}\}$. From $\overline{U} \subset \mathcal{O}$ it follows that

$$\Lambda \subset (\mu_0 - \varepsilon_0, \mu_0 + \varepsilon_0).$$

Fix $n \geq n_0$. From the continuity of the function $\delta(\cdot)$ in Definition 3.9 it follows that we can choose $0 < \varepsilon < \varepsilon_0$ and sufficiently small r

$$0 < r < \min\{\delta_{\mu_0-\varepsilon}, \delta_{\mu_0+\varepsilon}, \frac{\delta_{\mu_0-\varepsilon}}{\sqrt{T}}, \frac{\delta_{\mu_0+\varepsilon}}{\sqrt{T}}\}$$

such that $B_{r,\varepsilon}^{(n)} \subset U^{(n)}$ and

$$A_n x \notin \mathbf{P}_n \mathscr{P}_F(x, \mu) \quad \text{provided} \quad x \in B_C^{(n)}(0, r) \setminus \{0\}$$

for all $\mu \in [\mu_0 - \varepsilon_0, \mu_0 + \varepsilon_0] \setminus (\mu_0 - \varepsilon, \mu_0 + \varepsilon)$, where $B_C^{(n)}(0, r) = B_C(0, r) \cap C_T(I, H_n)$. From the choice of r, ε we have

$$\left\{ (x, \mu) : (x, \mu) \in \overline{U}^{(n)}, \ 0 \in G_n^r(x, \mu) \right\} \subset B_{r,\varepsilon}^{(n)}.$$

So, we obtain

$$deg(G_n^r, B_{r,\varepsilon}^{(n)}) = deg(G_n^r, \overline{U}^{(n)}) = 0.$$

Therefore, $ind V^{\sharp}(0, \mu_0) = 0$, giving a contradiction.
Thus, $\mathscr{W}$ is a non-compact component, i.e., either $\mathscr{W}$ is unbounded or $\overline{\mathscr{W}} \cap \overline{\mathcal{O}} \neq \emptyset$ and we obtain the conclusion of the theorem.   □

3.3.3   Application 3: Ordinary Feedback Control Systems in a Hilbert Space

In this section we consider the existence of periodic solutions for a feedback control problem in a Hilbert space.

Let $Y = C[0, 1]$ and $H = W^{1,2}[0, 1]$. It is clear that H is compactly embedded into Y and for every $y \in H$ we have: $\|y\|_Y \leq \|y\|_H$. Consider the periodic problem for a one-parameter family of the following feedback control systems in H:

$$\begin{cases} x'(t) = \mu a x(t) + f(x(t), u(t), \mu) \text{ for a.a. } t \in [0, T], \\ u(t) \in U\big(x(t), \mu\big) \text{ for a.a. } t \in [0, T], \\ x(0) = x(T), \end{cases} \tag{3.24}$$

where $a > 0$, $\mu \in \mathbf{R}$, the feedback multimap $U : Y \times \mathbf{R} \to P(Y)$ is u.s.c. and a map $f : Y \times Y \times \mathbf{R} \to H$ is continuous.

A function $x : [0, T] \to H$ satisfying (3.24) is called a trajectory of the system corresponding to the value μ of parameter, whereas a function $u : [0, T] \to Y$ is called the control.

We assume that the following conditions hold:

$(f1)$ There exist $c > 0$ and $0 < b < \frac{a}{2}$ such that

$$\|f(\varphi, \psi, \mu)\|_H \leq \|\varphi\|_Y \big(b|\mu| + c\|\psi\|_Y\big)$$

for all $(\varphi, \psi, \mu) \in Y \times Y \times \mathbf{R}$.

$(U1)$ For every $(\varphi, \mu) \in Y \times \mathbf{R}$ the set $f(\varphi, U(\varphi, \mu), \mu) \subset H$ is compact and convex.

$(U2)$ There exist $0 < M < \frac{a}{2c}$ such that

$$\|U(\varphi, \mu)\|_Y \le M(|\mu| + \|\varphi\|_Y)$$

for all $\varphi \in Y$, where c is the constant from $(f1)$.

Define a multimap $\tilde{F}: Y \times \mathbf{R} \to P(Y)$ by

$$\tilde{F}(\varphi, \mu) = \mu a \varphi + f(\varphi, U(\varphi, \mu), \mu).$$

We reduce the problem of global bifurcation of trajectories of (3.24) to the global bifurcation problem for the following family of inclusions:

$$\begin{cases} x'(t) \in F(x(t), \mu), & \text{for a.a. } t \in I, \\ x(0) = x(T), \end{cases} \tag{3.25}$$

where $F : H \times \mathbf{R} \to Kv(H)$ is the restriction of $\tilde{F}$.

It should be mentioned that, due to the classical Filippov Implicit Function Lemma (see, e.g., [25, 80]), each function x from the solution (x, μ) to (3.25) can be realized as the trajectory of the system (3.24) by the choice of the corresponding control, so problems (3.24) and (3.25) are equivalent.

Let us denote by $\mathscr{S}$ the set of all nontrivial trajectories of (3.24).

Theorem 3.10. *Let condition* $(f1)$ *and* $(U1)$–$(U2)$ *hold. Then there is a connected subset* $\mathscr{R} \subset \mathscr{S}$ *such that* $(0, 0) \in \overline{\mathscr{R}}$ *and* $\mathscr{R}$ *is unbounded.*

Proof. We prove that the problem (3.25) satisfies all condition of Theorem 3.9. At first, from $(f1)$ and $(U1) - (U2)$ it follows that $f(0, \psi, \mu) = 0$ for all $(\psi, \mu) \in Y \times \mathbf{R}$ and multimap F satisfies conditions $(H1) - (H2)$. Moreover, it is easy to verify that multimap $\tilde{F}$ satisfies all conditions of approximation solvability as given in Theorem 3.2. So, condition $(H3)$ holds.

We show that the functional

$$V_\mu: H \to \mathbf{R},$$

$$V_\mu(y) = \frac{1}{2}\mu\langle y, y\rangle_H$$

is a local integral guiding function at $(0, 0)$ for problem (3.25).

In fact, it is clear that V_μ is projectively homogeneous potential. Take $x \in W_T^{1,2}(I, H)$ and choose an arbitrary $g \in \mathscr{P}_F(x, \mu)$. Then, by the Filippov Implicit Function Lemma, there exists $u \in L^2(I, Y)$ such that $u(s) \in U(x(s))$ for a.e. $s \in I$ and

$$g(s) = \mu a x(s) + f(x(s), u(s), \mu) \text{ for a.e. } s \in I.$$

For $\mu \neq 0$, $\|x\|_2 \neq 0$ we have

$$\int_0^T \langle \nabla V_\mu(x(t)), g(t)\rangle_H dt = \int_0^T \langle \mu x(t), \mu a x(t) + f(x(t), u(t), \mu)\rangle_H dt$$

$$\geq a\mu^2 \|x\|_2^2 - |\mu| \int_0^T \|x(t)\|_H \|f(x(t), u(t), \mu)\|_H dt$$

$$\geq a\mu^2 \|x\|_2^2 - b\mu^2 \|x\|_2^2 - c|\mu| \int_0^T \|x(t)\|_H \|x(t)\|_Y \|u(t)\|_Y dt$$

$$\geq (a - b - Mc)\mu^2 \|x\|_2^2 - c|\mu| \int_0^T \|x(t)\|_H^2 M \|x(t)\|_Y dt$$

$$\geq \|x\|_2^2 |\mu| \Big((a - b - Mc)|\mu| - Mc\|x\|_C \Big) > 0 \qquad (3.26)$$

provided $|\mu| > 0$ and sufficiently small $\|x\|_C \neq 0$, where $\|x\|_C = \max_{[0,T]} \|x(t)\|_H$.

Thus V_μ is a local integral guiding function for problem (3.25) at the point $(0, 0)$. Now for sufficiently small $r, \varepsilon > 0$ consider the vector fields

$$V_n^\sharp : \overline{U}_{r,\varepsilon}^{(n)} \to \mathbf{R}^n \times \mathbf{R}, \ n \geq 1,$$

$$V_n^\sharp(y, \mu) = \{-Pr_n \nabla V_\mu(y), \varepsilon^2 - \mu^2\} = \{-\mu y, \varepsilon^2 - \mu^2\},$$

It is easy to see that

$$ind \ V_n^\sharp(0, 0) = 1 - (-1)^n = \begin{cases} 2 \text{ if } n \text{ is an odd number} \\ 0 \text{ if } n \text{ is an even number.} \end{cases}$$

Hence, $ind \ V_n^\sharp(0, 0) \neq 0$.

Moreover, from (3.26) it follows that for every $\mu \neq 0$ inclusion (3.25) has no solution x provided

$$0 < \|x\|_C < \frac{(a - b - Mc)|\mu|}{Mc}.$$

Therefore, $(0, 0)$ is the unique bifurcation point of inclusion (3.25). Applying Theorem 3.9 we obtain that there is a connected subset $\mathscr{R} \subset \mathscr{S}$ such that $(0, 0) \in \mathscr{R}$ and $\mathscr{R}$ is unbounded. $\square$

Chapter 4
Second-Order Differential Inclusions

Various aspects of the theory of second-order differential inclusions attract the attention of many researchers (see., e.g., [1, 2, 6, 12, 18, 42, 46, 47, 68, 70, 97]). In this chapter we consider the boundary value problem of form

$$u'' \in Q(u), \quad u(0) = u(1) = 0, \tag{4.1}$$

for second-order differential inclusions which arises naturally from some physical and control problems. Using the method of guiding functions we study the existence of solutions of problem (4.1) in an one-dimensional and in Hilbert spaces.

4.1 Existence Theorem in an One-Dimensional Space

By $W_0^{2,2}[0, 1]$ we denote the subset of $W^{2,2}[0, 1]$ consisting of all functions vanishing at the end-points of $[0, 1]$, i.e.,

$$W_0^{2,2}[0, 1] = \{u \in W^{2,2}[0, 1] : u(0) = u(1) = 0\}.$$

Define the continuous integral operator $j : L^2[0, 1] \to C[0, 1]$ by

$$(jf)(t) = \int_0^1 G(t, s) f(s) ds,$$

where

$$G(t, s) = \begin{cases} t(s-1) & \text{if } 0 \le t \le s, \\ s(t-1) & \text{if } s \le t \le 1. \end{cases}$$

Notice that the operator j in fact acts into $W_0^{2,2}[0, 1]$ and, for any $f \in L^2[0, 1]$, the boundary value problem

V. Obukhovskii et al., *Method of Guiding Functions in Problems of Nonlinear Analysis*, Lecture Notes in Mathematics 2076, DOI 10.1007/978-3-642-37070-0_4, © Springer-Verlag Berlin Heidelberg 2013

$$\begin{cases} u''(t) = f(t) \text{ for a.e. } t \in [0,1], \\ u(0) = u(1) = 0 \end{cases}$$

can be written in the form: $u = jf$ (see, e.g., [72]). By applying the Arzela–Ascoli theorem, it is easy to see also that the operator j transforms bounded sets into a relatively compact ones.

In this section we consider the existence of solutions to the following boundary value problem for the operator-differential inclusion

$$\begin{cases} u'' \in Q(u), \\ u(0) = u(1) = 0, \end{cases} \tag{4.2}$$

where $Q \colon C[0,1] \to C(L^2[0,1])$ is a multimap satisfying the following conditions:

(Q1) The composition $j \circ Q$ belongs to the class $CJ\big(C[0,1]; C[0,1]\big)$.
(Q2) There are constants $p, q > 0$ such that

$$\|Q(u)\|_2 \le q(1 + \|u\|_2^p)$$

for all $u \in C[0,1]$, where

$$\|Q(u)\|_2 = \sup\{\|f\|_2 \colon f \in Q(u)\}.$$

By a solution to problem (4.2) we mean a function $u \in W_0^{2,2}[0,1]$ such that there is a function $f \in Q(u)$ satisfying

$$u''(t) = f(t) \text{ for a.a. } t \in [0,1].$$

Remark 4.1. Let us mention that the class of multimaps Q satisfying condition (Q1) is large enough. For example, for every CJ-multimap Q the multimap $j \circ Q$ is a CJ-multimap. Moreover, there are multimaps Q which are not CJ-multimaps while $j \circ Q$ are CJ-multimaps. For example, let $F \colon [0,1] \times \mathbf{R} \to Kv(\mathbf{R})$ be a L^2-upper Carathéodory multimap. It is well known that the superposition multioperator $\mathscr{P}_F$ is well-defined, it is closed and has convex closed values.

Set $Q \colon C[0,1] \to Cv(L^2[0,1])$, $Q(x) = \mathscr{P}_F(x)$. From Proposition 1.17 it follows that the multimap $j \circ Q$ is closed. It is clear that for every bounded subset $U \subset C[0,1]$ the set $Q(U)$ is bounded in $L^2[0,1]$, therefore the set $j(Q(U))$ is a relatively compact set in $C[0,1]$. Hence, $j \circ Q$ is an u.s.c. multimap with compact convex values and so, it belongs to the class $J\big(C[0,1]; C[0,1]\big) \subset CJ\big(C[0,1]; C[0,1]\big)$.

The main result of this section is the following assertion.

Theorem 4.1. *Let conditions* (Q1) − (Q2) *hold. Assume that there exists $N > 0$ such that for every $u \in C[0,1]$, $\|u\|_2 > N$, the following relation holds*

$$\langle u, f \rangle_{L^2} = \int_0^1 u(s)f(s)ds > 0 \ \text{for all} \ f \in Q(u).$$

Then problem (4.2) has a solution.

Proof.   Problem (4.2) can be substituted by the following inclusion

$$u \in j \circ Q(u).$$

Condition $(Q2)$ implies that the set $Q(\Omega)$ is bounded in $L^2[0, 1]$ for every bounded subset $\Omega \subset C[0, 1]$. Therefore, the set $j \circ Q(\Omega)$ is relatively compact in $C[0, 1]$. Thus, $j \circ Q$ is a completely u.s.c. CJ-multimap.

Assume that there exists $u_* \in C[0, 1]$, such that $u_* \in j \circ Q(u_*)$. Notice that then $u_*(0) = u_*(1) = 0$. Then there is $f_* \in Q(u_*)$ such that $u_*''(t) = f_*(t)$ for a.a. $t \in [0, 1]$, and hence

$$\langle f_*, u_* \rangle_{L^2} = \langle u_*'', u_* \rangle_{L^2} = -\langle u_*', u_*' \rangle_{L^2} \le 0.$$

Therefore, $\|u_*\|_2 \le N$.
For every $t \in [0, 1]$, we have

$$|u_*(t)| \le \int_0^1 |G(t, s)||f_*(s)|ds \le \int_0^1 |f_*(s)|ds \le \|f_*\|_2.$$

From $(Q2)$ it follows that for every $t \in [0, 1]$

$$|u_*(t)| \le \|f_*\|_2 \le \|Q(u_*)\|_2 \le q(1 + N^p),$$

hence, $\|u_*\|_C = \max_{[0,1]} |u(t)| \le q(1 + N^p)$.
Now set $R = qN^p + q + 1$. Consider the multimap

$$\Psi \colon B_C(0, R) \times [0, 1] \to K(C[0, 1]),$$

$$\Psi(u, \lambda) = j \circ \big((1 - \lambda)\delta u + \lambda Q(u)\big),$$

where $0 < \delta < \frac{1}{N}$ is an arbitrary number.

Let us show that Ψ is a compact CJ-multimap. In fact, from condition $(Q1)$ it follows that we can represent the multimap $j \circ Q$ as $\varphi \circ F \in CJ(C[0, 1]; C[0, 1])$, where $F \colon C[0, 1] \to K(Y)$ is a J-multimap from $C[0, 1]$ to some metric space Y and $\varphi \colon Y \to C[0, 1]$ is a continuous map. Define the multimap

$$\tilde{F} \colon B_C(0, R) \times [0, 1] \to K(C[0, 1] \times Y \times \mathbf{R}),$$

$$\tilde{F}(u, \lambda) = \{u\} \times F(u) \times \{\lambda\},$$

and the map

$$\tilde{\varphi} : C[0, 1] \times Y \times \mathbf{R} \to K(C[0, 1]),$$

$$\tilde{\varphi}(u, v, \lambda) = \delta(1 - \lambda)ju + \lambda\varphi(v).$$

It is clear that $\tilde{F}$ is a J-multimap, $\tilde{\varphi}$ is a continuous map and for every $(u, \lambda) \in B_C(0, R) \times [0, 1]$ we have $\Psi(u, \lambda) = \tilde{\varphi} \circ \tilde{F}(u, \lambda)$. So, Ψ is a CJ-multimap. Further, the sets $j \circ Q(B_C(0, R))$ and $j(B_C(0, R))$ are relatively compact in $C[0, 1]$, therefore $\Psi(B_C(0, R) \times [0, 1])$ is a relatively compact set in $C[0, 1]$, too. So, the multimap Ψ is compact.

Now, we prove that Ψ has no fixed points on $\partial B_C(0, R) \times [0, 1]$. To the contrary, assume that there exists $(u_*, \lambda_*) \in \partial B_C(0, R) \times [0, 1]$ such that $u_* \in \Psi(u_*, \lambda_*)$. Then there is a function $f_* \in Q(u_*)$ such that

$$u_*(t) = \int_0^1 G(t, s)\Big((1 - \lambda_*)\delta u_*(s) + \lambda_* f_*(s)\Big)ds, \; \forall t \in [0, 1], \tag{4.3}$$

or equivalently,

$$\begin{cases} u_*''(t) = (1 - \lambda_*)\delta u_*(t) + \lambda_* f_*(t), \; \text{for a.a. } t \in [0, 1], \\ u_*(0) = u_*(1) = 0. \end{cases} \tag{4.4}$$

Assume that $\|u_*\|_2 \leq N$. Then from (4.3) we have

$$|u_*(t)| \leq \delta(1 - \lambda_*)\|u_*\|_2 + \lambda_*\|f_*\|_2 \leq \delta(1 - \lambda_*)N + \lambda_* q(1 + N^p) < R,$$

for all $t \in [0, 1]$. Hence $u_* \notin \partial B_C(0, R)$, that is the contradiction. Therefore, $\|u_*\|_2 > N$. From (4.4) it follows that

$$\langle u_*'', u_*\rangle_{L^2} = \delta(1 - \lambda_*)\langle u_*, u_*\rangle_{L^2} + \lambda_* \langle u_*, f_*\rangle_{L^2} > 0,$$

giving a contradiction.

Thus, Ψ is a homotopy joining $\Psi(\cdot, 0) = \delta j \circ i$ and $\Psi(\cdot, 1) = j \circ Q$, where i denotes the inclusion map. The homotopic invariance property of the topological degree implies that

$$deg(i - j \circ Q, B_C(0, R)) = deg(i - \delta j \circ i, B_C(0, R)).$$

For a sufficiently small $\delta > 0$ we have

$$\|u - (u - \delta ju)\|_C = \delta\|ju\|_C < \|u\|_C$$

for all $u \in \partial B_C(0, R)$.

Then the vector fields i and $i - \delta j \circ i$ are homotopic on $\partial B_C(0, R)$ (see Lemma 1.4), so

$$deg(i - j \circ Q, B_C(0, R)) = deg(i - \delta j \circ i, B_C(0, R)) = deg(i, B_C(0, R)) = 1.$$

Hence problem (4.2) has a solution $u \in B_C(0, R)$. $\square$

Now we can formulate the above result in terms of the guiding functions. Notice that for every continuous function $V: \mathbf{R} \to \mathbf{R}$ the following map

$$V^\sharp: C[0, 1] \to L^2[0, 1], \quad V^\sharp(u)(t) = V(u(t)),$$

is continuous.

Definition 4.1. A continuous function $V: \mathbf{R} \to \mathbf{R}$ is said to be an integral guiding function for problem (4.2), if:

$(V1)$ there are $\alpha \geq 0$ and $\beta > 0$ such that

$$|V(t)| \leq \alpha + \beta|t|, \quad \forall t \in \mathbf{R};$$

$(V2)$ there exists $N > 0$ such that for every $u \in C[0, 1]$, $\|u\|_2 > N$, the following relation holds:
$$\langle V^\sharp(u), f \rangle_{L^2} > 0 \text{ for all } f \in Q(u),$$

$(V3)$ for every $u \in W_0^{2,2}[0, 1]$, from $\|u\|_2 > N$ it follows that

$$\langle u'', V^\sharp(u) \rangle_{L^2} \leq 0,$$

where N is the same constant as in $(V2)$.

Theorem 4.2. *Let conditions $(Q1)$–$(Q2)$ hold. Assume that there exists an integral guiding function V for problem (4.2). Then problem (4.2) has a solution.*

Proof. Set $R = qN^p + q + 1$ and consider the multimap

$$\Psi: B_C(0, R) \times [0, 1] \to K(C[0, 1]),$$

$$\Psi(u, \lambda) = j \circ \big((1 - \lambda)\delta V^\sharp(u) + \lambda Q(u)\big),$$

where $\delta, 0 < \delta < \frac{1}{\alpha + \beta N}$ is an arbitrary number, with N being the number in $(V2)$. In a similar way to the proof of Theorem 4.1, one can verify that Ψ is a CJ-multimap. From $(Q2)$ and $(V1)$ it follows that the sets $Q\big(B_C(0, R)\big)$ and $V^\sharp\big(B_C(0, R)\big)$ are bounded in $L^2[0, 1]$. Since the operator j is completely continuous, the sets $j \circ Q\big(B_C(0, R)\big)$ and $j \circ V^\sharp\big(B_C(0, R)\big)$ are relatively compact in $C[0, 1]$. Hence the set $\Psi\big(B_C(0, R) \times [0, 1]\big)$ is relatively compact in $C[0, 1]$. Thus, Ψ is a compact CJ-multimap.

Let us show that Ψ has no fixed points on $\partial B_C(0, R) \times [0, 1]$. To the contrary, assume that there is $(u_*, \lambda_*) \in \partial B_C(0, R) \times [0, 1]$ such that $u_* \in \Psi(u_*, \lambda_*)$. Then there is a function $f_* \in Q(u_*)$ such that

$$u_*(t) = \int_0^1 G(t,s)\Big((1-\lambda_*)\delta V(u_*(s)) + \lambda_* f_*(s)\Big)ds, \ \forall t \in [0,1], \qquad (4.5)$$

or equivalently,

$$\begin{cases} u_*''(t) = (1-\lambda_*)\delta V(u_*(t)) + \lambda_* f_*(t), \ for \ a.e. \ t \in [0,1], \\ u_*(0) = u_*(1) = 0. \end{cases} \qquad (4.6)$$

Assume that $\|u_*\|_2 \le N$. Then from $(Q2)$, $(V1)$ and (4.5) we have

$$|u_*(t)| \le \delta(1-\lambda_*)\int_0^1 |V(u_*(t))|\,dt + \lambda_* \int_0^1 |f_*(t)|\,dt$$

$$\le \delta(1-\lambda_*)\int_0^1 (\alpha + \beta|u_*(t)|)\,dt + \lambda_*\|f_*\|_2$$

$$\le \delta(1-\lambda_*)(\alpha + \beta\|u_*\|_2) + \lambda_* q(1+N^p) \le \delta(\alpha + \beta N) + q(1+N^p) < R,$$

for all $t \in [0,1]$.

Hence $u_* \notin \partial B_C(0,R)$, that is the contradiction. Therefore, $\|u_*\|_2 > N$. Notice that $u_* \in W_0^{2,2}[0,1] \subset C[0,1]$, then from $(V2)$–$(V3)$ and (4.6) it follows that

$$\langle u_*'', V^\sharp(u_*)\rangle_{L^2} = \delta(1-\lambda_*)\langle V^\sharp(u_*), V^\sharp(u_*)\rangle_{L^2} + \lambda_*\langle V^\sharp(u_*), f_*\rangle_{L^2} > 0,$$

giving a contradiction.

So, we again obtain that Ψ is a homotopy and for sufficiently small $\delta > 0$ we have

$$deg(i - j \circ Q, B_C(0,R)) = deg(i - \delta j \circ V^\sharp, B_C(0,R)) = deg(i, B_C(0,R)) = 1.$$

Thus problem (4.2) has a solution.   □

4.2 Applications

4.2.1 Equations with Discontinuous Nonlinearities

In this section we consider the following equation

$$\begin{cases} -u''(t) + g(u(t)) = \varphi(t, u(t)), \\ u(0) = u(1) = 0. \end{cases} \qquad (4.7)$$

where the function $g: \mathbf{R} \to \mathbf{R}$ and the function $\varphi: [0,1] \times \mathbf{R} \to \mathbf{R}$ satisfy the following conditions:

($g1$) g is continuous and there is $a > 0$ such that

$$|g(x)| \leq a(1 + |x|),$$

for all $x \in \mathbf{R}$.

($\varphi1$) For a.e. $t \in [0, 1]$ there exist finite limits

$$\underline{\varphi}(t, \xi) = \liminf_{\xi' \to \xi} \varphi(t, \xi'); \qquad \overline{\varphi}(t, \xi) = \limsup_{\xi' \to \xi} \varphi(t, \xi')$$

and the functions $\underline{\varphi}$, $\overline{\varphi}$ are superpositionally measurable, i.e. the functions $\underline{\varphi}(t, \psi(t)), \overline{\varphi}(t, \psi(t))$ are measurable for each measurable function $t \to \psi(t)$.

($\varphi2$) There exist functions $f_*, f^* \in L^2[0, 1]$ such that

$$f_*(t) \leq \varphi(t, \xi) \leq f^*(t)$$

for a.e. $t \in [0, 1]$ and all $\xi \in \mathbf{R}$.

Let us recall (see, e.g. [93]) that Carathéodory functions, pointwise limits of continuous functions, and Borel measurable functions belong to the class of superpositionally measurable functions.

Denote by $[f_*, f^*] \subset L^2[0, 1]$ the interval

$$[f_*, f^*] = \{ f \in L^2[0, 1] : f_*(t) \leq f(t) \leq f^*(t) \text{ for } a.a. \ t \in [0, 1] \}$$

and define the multimap $\Phi : C[0, 1] \to Cv(L^2[0, 1])$ by the rule

$$\Phi(u) = \left[\underline{\varphi}(x, u(x)), \overline{\varphi}(x, u(x)) \right].$$

According to [32], Theorem 1.1 let us mention that the multimap Φ is u.s.c.

So, we can substitute the family of (4.7) by the following operator inclusion

$$u'' \in \tilde{g}(u) - \Phi(u),$$

whose solutions are called the generalized solutions to (4.7), where $\tilde{g} : C[0, 1] \to C[0, 1]$,

$$\tilde{g}(u)(t) = g(u(t)), \ t \in [0, 1].$$

It is easy to see that the multimap $Q(u) = \tilde{g}(u) - \Phi(u)$ satisfies conditions ($Q1$)–($Q2$). Hence, by virtue of Theorem 4.1 we obtain the sufficient conditions for existence of solutions to problem (4.7).

In turn, let us mention that equations of type (4.7) appear in many problems of mathematical physics. For example, Lavrentiev's problem on detachable currents at the presence of nonlinear perturbations in one-dimensional space can be described by the following equation (cf. [94]):

$$-u''(t) + g(u(t)) = \mu \, sign(u(t)),$$
$$u(0) = u(1) = 0,$$

where $\mu > 0$.

Here

$$\underline{\varphi}(t, \xi) = \begin{cases} \mu, & \xi > 0, \\ -\mu, & \xi \leq 0; \end{cases}$$

and

$$\overline{\varphi}(t, \xi) = \begin{cases} \mu, & \xi \geq 0, \\ -\mu, & \xi < 0. \end{cases}$$

Let us mention that (4.7) has the form of an elliptic equations as considered in [94] for the case when the nonlinearity $\varphi - g$ is "strongly" bounded, i.e.,

$$|\varphi(t, x) - g(x)| \leq \beta < \infty$$

for all $x \in \mathbf{R}$ and a.a. $t \in [0, 1]$. So here we get an extended result. Moreover, (4.7) can be represented as

$$Au(x) + h(u)(x) = \varphi(x, u(x)), \tag{4.8}$$

where $Au = -u'' - u$, $h(u) = \tilde{g}(u) + u$. It is clear that A is a linear Fredholm operator of index zero. The sufficient conditions for the existence of a solution of the equation containing a linear Fredholm operator of index zero and a discontinuous nonlinearity of form (4.8) are given in [119] when the map h satisfies the following condition: there exist finite limits

$$h(-\infty) = \lim_{r \to -\infty} h(r); \qquad h(+\infty) = \lim_{r \to +\infty} h(r),$$

and

$$h(-\infty) \leq h(r) \leq h(+\infty)$$

for all $r \in \mathbf{R}$. So, in general that couldn't be applied the results of [116] to our situation. Now let us take an illustrative illustrating.

Example 4.1. Consider the following equation

$$\begin{cases} -u''(t) + (\mu^2 + 1)u(t) + \mu + 1 = \mu \, sign(u(t)), \ \mu > 0, \\ u(0) = u(1) = 0, \end{cases} \tag{4.9}$$

or equivalently,

$$\begin{cases} u'' \in Q(u) = g(u) - \Phi(u), \\ u(0) = u(1) = 0, \end{cases}$$

where Φ is defined as above and $g(u) = (\mu^2 + 1)u + \mu + 1$.

For every $f \in Q(u)$ there exists $\omega \in \Phi(u)$ such that

$$f = (\mu^2 + 1)u + \mu + 1 - \omega.$$

Notice that $\omega \in [-\mu, \mu]$. We have that

$$\begin{aligned}
\langle f, u \rangle_{L^2} &= \langle (\mu^2 + 1)u + \mu + 1 - \omega, u \rangle_{L^2} \\
&\geq (\mu^2 + 1)\|u\|_2^2 - (\mu + 1)\|u\|_2 - \|\omega\|_2\|u\|_2 \\
&\geq (\mu^2 + 1)\|u\|_2^2 - (2\mu + 1)\|u\|_2 > 0
\end{aligned}$$

provided $\|u\|_2 > \frac{2\mu+1}{\mu^2+1}$.

By virtue of Theorem 4.1 the problem (4.9) has a solution.

4.2.2 Boundary Value Problem

Now we consider a more general class of differential equations with discontinuities that is

$$\begin{cases} u''(t) - \lambda u \in F(t, u(t)), \\ u(0) = u(1) = 0, \end{cases} \tag{4.10}$$

where $\lambda > 0$ and $F : [0, 1] \times \mathbf{R} \to Kv(\mathbf{R})$ is a L^2-upper Carathéodory multimap satisfying condition:

$$\|F(t, x)\| \leq K(1 + |x|), \text{ for all } x \in \mathbf{R} \text{ and a.e. } t \in [0, 1],$$

for some $K > 0$.

Theorem 4.3. *For each $\lambda > K$ problem (4.10) has a solution.*

Proof. Problem (4.10) can be substituted as inclusion (4.2), where $Q(u) = \lambda u + \mathscr{P}_F(u)$. It is easy to verify that the multimap Q satisfies conditions $(Q1)$–$(Q2)$.

Now choosing $\lambda > K$, for every $f \in Q(u)$ there is $\omega \in \mathscr{P}_F(u)$ such that $f = \lambda u + \omega$. We have that

$$\begin{aligned}
\langle f, u \rangle_{L^2} &= \lambda \langle u, u \rangle_{L^2} + \langle \omega, u \rangle_{L^2} \\
&\geq \lambda \|u\|_2^2 - \int_0^1 K|u(s)|(1 + |u(s)|)ds \\
&\geq (\lambda - K)\|u\|_2^2 - K\|u\|_2 > 0,
\end{aligned}$$

provided $\|u\|_2 > \frac{K}{\lambda - K}$. By virtue of Theorem 4.1 problem (4.10) has a solution. $\square$

4.2.3   A Second-Order Differential Equation

Consider the following equation

$$v''(t) = \frac{g(t)}{(v^2(t) + 1)^{\mu}} - \frac{h(t)}{(v^2(t) + 1)^{\lambda}} + cv(t) + d, \text{ for a.a. } t \in [0, 1], \quad (4.11)$$

$$v(0) = v(1) = 0$$

where $c, \mu, \lambda > 0$; $d \in \mathbf{R}$ and $g, h \in L^2[0, 1]$.

Theorem 4.4. *Equation (4.11) has a solution.*

Proof. Problem (4.11) can be substituted by

$$v = j \circ \gamma(v),$$

where $\gamma : C[0, 1] \to L^2[0, 1]$,

$$\gamma(v)(t) = \frac{g(t)}{(v^2(t) + 1)^{\mu}} - \frac{h(t)}{(v^2(t) + 1)^{\lambda}} + cv(t) + d.$$

The operator γ satisfies conditions $(Q1)$–$(Q2)$ and

$$\langle v, \gamma(v) \rangle_{L^2} = \int_0^1 \frac{g(t)v(t)}{(v^2(t) + 1)^{\mu}} dt - \int_0^1 \frac{h(t)v(t)}{(v^2(t) + 1)^{\lambda}} dt + \int_0^1 v(t)(cv(t) + d)dt$$

$$\geq c\|v\|_2^2 - |d| \int_0^1 |v(s)| ds - \int_0^1 \frac{|g(t)v(t)|}{(v^2(t) + 1)^{\mu}} dt - \int_0^1 \frac{|h(t)v(t)|}{(v^2(t) + 1)^{\lambda}} dt$$

$$\geq c\|v\|_2^2 - (|d| + \|g\|_2 + \|h\|_2)\|v\|_2 > 0,$$

provided

$$\|v\|_2 > \frac{|d| + \|g\|_2 + \|h\|_2}{c}.$$

From Theorem 4.1 it follows that (4.11) has a solution.   □

4.2.4   Feedback Control Systems

Consider a feedback control system of the form

$$\begin{cases} u''(t) - \lambda u(t) = f(t, u(t), v(t)), \text{ for a.a. } t \in [0, 1], \\ v'(t) \in G(t, v(t), u(t)), \text{ for a.a. } t \in [0, 1], \\ u(0) = u(1) = 0, v(0) = v_0, \end{cases} \quad (4.12)$$

where $v_0 \in \mathbf{R}$, $\lambda > 0$, $f : [0, 1] \times \mathbf{R} \times \mathbf{R} \to \mathbf{R}$ is an upper Carathéodory map and $G : [0, 1] \times \mathbf{R} \times \mathbf{R} \to Kv(\mathbf{R})$ is an upper Carathéodory multimap.

Assume that

(A1) There is $\alpha > 0$ such that

$$|f(t, x, y)| \le \alpha(1 + |x| + |y|)$$

for all $(x, y) \in \mathbf{R} \times \mathbf{R}$ and a.e. $t \in [0, 1]$.

(A2) The multimap G is uniformly continuous with respect to the third argument in the sense: for every $\varepsilon > 0$ there is $\delta > 0$ such that

$$G(t, x, \bar{y}) \subset O_\varepsilon(G(t, x, y)) \quad \forall (t, x) \in [0, 1] \times \mathbf{R}$$

whenever $|\bar{y} - y| < \delta$.

(A3) There is $\beta > 0$ such that

$$\|G(t, x, y)\| = \max\{|z| : z \in G(t, x, y)\} \le \beta(1 + |x| + |y|)$$

for all $(x, y) \in \mathbf{R} \times \mathbf{R}$ and a.e. $t \in [0, 1]$.

For a given function $u \in C[0, 1]$ define the multimap

$$G_u : [0, 1] \times \mathbf{R} \to Kv(\mathbf{R}), \quad G_u(t, x) = G(t, x, u(t)).$$

Using Proposition 1.16 and the above conditions we conclude that the multimap $G_u(t, x)$ is upper Carathéodory. From Proposition 2.1 and the continuous dependence of the solution set of a differential inclusion on a parameter (see, e.g. [80]) we know that:

for each $u \in C[0, 1]$ the set Φ_u of all solutions of the following problem

$$\begin{cases} v'(t) \in G(t, v(t), u(t)) \ for \ a.a. \ t \in [0, 1] \\ v(0) = v_0 \end{cases}$$

is an R_δ-set in $C[0, 1]$ and the multimap

$$\Phi : C[0, 1] \to K(C[0, 1]), \quad \Phi(u) = \Phi_u,$$

is upper semicontinuous.

By a solution to problem (4.12) we mean a function $u \in W_0^{2,2}[0, 1]$ such that there is an absolutely continuous function $v \in \Phi(u)$ satisfying

$$u''(t) - \lambda u(t) = f(t, u(t), v(t)), \ for \ a.a. \ t \in [0, 1].$$

Theorem 4.5. *Let conditions (A1)–(A3) hold. Then for each*

$$\lambda > \alpha(1 + \beta e^\beta)$$

the feedback control system (4.12) has a solution.

Proof. Let $\tilde{\Phi}: C[0,1] \to K(C[0,1] \times C[0,1])$

$$\tilde{\Phi}(u) = \{u\} \times \Phi(u),$$

and $\tilde{f}: C[0,1] \times C[0,1] \to L^2[0,1]$,

$$\tilde{f}(u,v)(t) = \lambda u(t) + f(t,u(t),v(t)), \ t \in [0,1].$$

Then we can substitute the feedback control system (4.12) by the following problem

$$\begin{cases} u'' \in Q(u), \\ u(0) = u(1) = 0, \end{cases} \qquad (4.13)$$

where $Q: C[0,1] \to K(L^2[0,1])$,

$$Q(u) = \tilde{f} \circ \tilde{\Phi}(u).$$

From the continuity of the operator $\tilde{f}$ and the fact that $\tilde{\Phi} \in J(C[0,1]; C[0,1] \times C[0,1])$ it follows that $j \circ Q \in CJ(C[0,1]; C[0,1])$.

Now letting $g \in Q(u)$, there exists $v \in \Phi(u)$ such that

$$g(s) = \tilde{f}(u,v)(s) = \lambda u(s) + f(s,u(s),v(s)), \ \forall s \in [0,1].$$

From $v \in \Phi(u)$ it follows that there is a function $h \in L^1[0,1]$ such that

$$h(t) \in G(t,v(t),u(t)), \ for \ a.a. \ t \in [0,1],$$

and

$$v(t) = v_0 + \int_0^t h(s)ds, \ 0 \le t \le 1.$$

From (A3) it follows that for every $t \in [0,1]$ the following relations hold

$$|v(t)| \le |v_0| + \int_0^t |h(s)|ds \le |v_0| + \int_0^t \beta(1 + |v(s)| + |u(s)|)ds$$

$$\le |v_0| + \beta + \beta \|u\|_2 + \int_0^t \beta |v(s)|ds.$$

By Lemma 2.1 we obtain

$$|v(t)| \leq (|v_0| + \beta + \beta \|u\|_2) e^{\beta}, \quad \forall t \in [0, 1].$$

Applying $(A3)$ we have

$$
\begin{aligned}
\|g\|_2^2 &= \int_0^1 g^2(s)ds = \int_0^1 \Big(\lambda u(s) + f\big(s, u(s), v(s)\big) \Big)^2 ds \\
&\leq \int_0^1 (\lambda^2 + 1)\Big(u^2(s) + f^2(s, u(s), v(s)) \Big) ds \\
&\leq (\lambda^2 + 1)\Big(\|u\|_2^2 + \int_0^1 \alpha^2 (1 + |u(s)| + |v(s)|)^2 ds \Big) \\
&\leq (\lambda^2 + 1)\Big(\|u\|_2^2 + \int_0^1 3\alpha^2 (1 + u^2(s) + v^2(s)) ds \Big) \\
&\leq (\lambda^2 + 1)\Big((1 + 3\alpha^2)\|u\|_2^2 + 3\alpha^2 + 3\alpha^2 (|v_0| + \beta + \beta\|u\|_2)^2 e^{2\beta} \Big).
\end{aligned}
$$

Therefore, the multimap Q satisfies condition $(Q2)$.

Now for every $u \in C[0, 1]$, choosing an arbitrary $g \in Q(u)$, we have that

$$
\begin{aligned}
\langle g, u \rangle_{L^2} &= \int_0^1 u(s)\Big(\lambda u(s) + f\big(s, u(s), v(s)\big) \Big) ds \\
&\geq \lambda \|u\|_2^2 - \int_0^1 |f(s, u(s), v(s))| \, |u(s)| ds \\
&\geq \lambda \|u\|_2^2 - \alpha \int_0^1 |u(s)|(1 + |u(s)| + |v(s)|) ds \\
&\geq (\lambda - \alpha)\|u\|_2^2 - \alpha \int_0^1 |u(s)| ds - \alpha e^{\beta}(|v_0| + \beta + \beta\|u\|_2) \int_0^1 |u(s)| ds \\
&\geq (\lambda - \alpha - \alpha\beta e^{\beta})\|u\|_2^2 - \alpha(1 + e^{\beta}(|v_0| + \beta))\|u\|_2 > 0
\end{aligned}
$$

provided

$$\|u\|_2 > \frac{\alpha(1 + e^{\beta}(|v_0| + \beta))}{\lambda - \alpha(1 + \beta e^{\beta})}.$$

By virtue of Theorem 4.1 problem (4.13) has a solution, and hence the feedback control system (4.12) has a solution. □

4.2.5 A Model of a Motion of a Particle in a One-Dimensional Potential

It is well known that for a single particle in a one-dimensional potential energy V, the time-independent Schroedinger equation takes the form:

$$\frac{-\hbar^2}{2m} \Psi''(x) + V(x)\Psi(x) = E\Psi(x), \tag{4.14}$$

where m is the particle's mass, $\hbar$—the reduced Planck constant, E—the total energy of the particle, $V(x)$ is the potential energy at the position x and $\Psi(x)$ is the wave function.

Here we consider the case when the potential energy $V(x)$ has a form:

$$V(x) = \begin{cases} V, & 0 \le x \le 1, \\ 0, & \textit{otherwise,} \end{cases}$$

where V is a constant.

In this case, (4.14) must be solved in three regions:

$$I\ (x < 0),\ II\ (0 \le x \le 1)\ and\ III\ (x > 1).$$

The corresponding solutions of (4.14) in the first and third regions are

$$\Psi_I(x) = A \sin kx + B \cos kx \quad and \quad \Psi_{III}(x) = C \sin kx + D \cos kx,$$

where $k = \frac{\sqrt{2mE}}{\hbar}$ and A, B, C, D are constants.

So, we focus our attention on the solution of (4.14) in the second region. In this region the Schroedinger equation has the form:

$$\Psi_{II}''(x) = \frac{2m}{\hbar^2} V \Psi_{II}(x) - \frac{2m}{\hbar^2} E \Psi_{II}(x). \tag{4.15}$$

We assume that the potential V is connected with the wave function Ψ_{II} by the following relation:

$$V \in F(\Psi_{II}), \tag{4.16}$$

where $F: L^2[0, 1] \to K(\mathbf{R}_+)$ is a J-multimap, $\mathbf{R}_+ = [0, +\infty)$.

From the continuity of Ψ it follows that in region II the boundary conditions for (4.15) are:

$$\Psi_{II}(0) = \Psi_I(0) = B \quad and \quad \Psi_{II}(1) = \Psi_{III}(1) = C \sin k + D \cos k. \tag{4.17}$$

By a solution of problem (4.15)–(4.17) we mean a function $\Psi_{II} \in W^{2,2}[0, 1]$ for which there exists $V \in F(\Psi_{II})$ such that (4.15) and condition (4.17) hold.

Theorem 4.6. *Assume that there exist $a, b > 0$ such that*

$$F(u) \subseteq \left[a\|u\|_2, \; b(1 + \|u\|_2) \right] \text{ for all } u \in L^2[0, 1].$$

Then problem (4.15)–(4.17) has a solution.

Proof. Set $\alpha = B$, $\beta = C \sin k + D \cos k$ and $g(x) = \beta x + \alpha(1 - x)$. For every $x \in [0, 1]$ let $\varphi(x) = \Psi_{II}(x) - g(x)$. Then problem (4.15)–(4.17) can be replaced with the following system

$$\begin{cases} \varphi''(x) = \frac{2m}{\hbar^2} V(\varphi(x) + g(x)) - \frac{2m}{\hbar^2} E(\varphi(x) + g(x)), \\ V \in F(\varphi + g), \\ \varphi(0) = \varphi(1) = 0, \end{cases}$$

or equivalently,

$$\varphi \in j \circ Q(\varphi), \tag{4.18}$$

where

$$Q(\varphi) = \frac{2m}{\hbar^2}(\varphi + g)F(\varphi + g) - \frac{2m}{\hbar^2}E(\varphi + g),$$

and the operator j is defined as in Sect. 4.1.

It is easy to see that the multimap Q satisfies conditions (Q1)–(Q2). For every $w \in Q(\varphi)$ there is $V \in F(\varphi + g)$ such that

$$w = \frac{2m}{\hbar^2}(V - E)\varphi + \frac{2m}{\hbar^2}(V - E)g.$$

We have

$$\langle \varphi, w \rangle_{L^2} = \frac{2m}{\hbar^2}(V - E)\|\varphi\|_2^2 + \frac{2m}{\hbar^2}(V - E)\langle g, \varphi \rangle_{L^2}$$

$$\geq \frac{2m}{\hbar^2}(a\|\varphi + g\|_2 - E)\|\varphi\|_2^2 - \frac{2m}{\hbar^2}V\|g\|_2\|\varphi\|_2 - \frac{2m}{\hbar^2}E\|g\|_2\|\varphi\|_2$$

$$\geq \frac{2m}{\hbar^2}(a\|\varphi\|_2 - a\|g\|_2 - E)\|\varphi\|_2^2 - \frac{2m}{\hbar^2}(b + b\|\varphi\|_2 + b\|g\|_2)\|g\|_2\|\varphi\|_2$$

$$- \frac{2m}{\hbar^2}E\|g\|_2\|\varphi\|_2.$$

Therefore,

$$\langle \varphi, w \rangle_{L^2} \geq \frac{2m}{\hbar^2}a\|\varphi\|_2^3 - \frac{2m}{\hbar^2}(a\|g\|_2 + E + b\|g\|_2)\|\varphi\|_2^2$$

$$- \frac{2m}{\hbar^2}\|g\|_2(b + b\|g\|_2 + E)\|\varphi\|_2 > 0$$

for sufficiently large $\|\varphi\|_2$.

From Theorem 4.1 it follows that inclusion (4.18) is solvable, and hence problem
(4.15)–(4.17) has a solution.   □

4.3 Existence Theorem in Hilbert Spaces

In this section we consider boundary value problem (4.1) in an infinite-dimensional
case.

Let H be a real infinite-dimensional Hilbert space with an orthonormal basis
$\{e_n\}_{n=1}^{\infty}$. For every $n \in \mathbf{N}$, let H_n be a n-dimensional subspace H with basis
$\{e_k\}_{k=1}^{n}$ and P_n be the projection onto H_n. Set $I = [0,1]$. The symbol $\langle f, g \rangle_{L^2}$
denotes the inner product of elements $f, g \in L^2(I, H)$. We consider the Sobolev
space $W^{k,2}(I, H)$ and its subspace $W_0^{k,2}(I, H)$. Notice that for every $k \geq 1$ the
embedding $W^{k,2}(I, H) \hookrightarrow C(I, H)$ is continuous (but not compact). The weak
convergence in $W^{k,2}(I, H)$ $[L^2(I, H)]$ is denoted by $u_n \overset{W^{k,2}}{\rightharpoonup} u_0$ [resp., $f_n \overset{L^2}{\rightharpoonup} f_0$].

For every $n \in \mathbf{N}$ let $J_n : L^2(I, H) \to C(I, H_n)$ be the operator defined by

$$J_n(f)(t) = \sum_{k=1}^{n} \left(\int_0^1 G(t, s) f_{(k)}(s) ds \right) e_k,$$

where the function $G(t, s)$ is defined as in Sect. 4.1 and

$$f(t) = \sum_{k=1}^{\infty} f_{(k)}(t) e_k, \quad \text{for all } t \in I.$$

It is clear that the operator J_n is completely continuous and for each $t \in I$ we have

$$\begin{aligned}
\| J_n(f)(t) \|_H &= \left(\sum_{k=1}^{n} \left(\int_0^1 G(t, s) f_{(k)}(s) ds \right)^2 \right)^{\frac{1}{2}} \\
&\leq \left(\sum_{k=1}^{n} \int_0^1 G^2(t, s) ds \int_0^1 f_{(k)}^2(s) ds \right)^{\frac{1}{2}} \\
&\leq \left(\int_0^1 \sum_{k=1}^{n} f_{(k)}^2(s) ds \right)^{\frac{1}{2}} \leq \left(\int_0^1 \| f(s) \|_H^2 ds \right)^{\frac{1}{2}} = \| f \|_2.
\end{aligned}$$

$$(4.19)$$

For $n \in \mathbf{N}$, define the projection $\mathbf{P}_n : L^2(I, H) \to L^2(I, H_n)$ generated by P_n as

$$(\mathbf{P}_n f)(t) = P_n f(t), \quad \text{for a.a. } t \in I.$$

Consider now the following operator-differential inclusion

$$\begin{cases} u'' \in Q(u), \\ u(0) = u(1) = 0, \end{cases} \qquad (4.20)$$

where the multimap $Q: C(I, H) \rightarrow C(L^2(I, H))$ satisfies the following conditions:

$(Q1)'$ For each $m \in \mathbf{N}$ the restriction $(J_m \circ Q)|_{C(I, H_m)}$ belongs to $CJ\big(C(I, H_m)$; $C(I, H_m)\big)$.

$(Q2)'$ There are constants $p_1, q_1 > 0$ such that

$$\|Q(u)\|_2 \leq q_1(1 + \|u\|_2^{p_1})$$

for all $u \in C(I, H)$.

$(Q3)$ For every bounded closed subset $M \subset W_0^{2,2}(I, H)$, if there exist the sequences $\{n_k\}$ and $\{u_k\}$, $u_k \in M \cap W_0^{2,2}(I, H_{n_k})$ such that

$$u_k'' \in P_{n_k} Q(u_k),$$

then there is $u_* \in M$ such that $u_*'' \in Q(u_*)$.

By a solution to problem (4.20) we mean a function $u \in W_0^{2,2}(I, H)$ such that there is a function $f \in Q(u)$ and

$$u''(t) = f(t) \text{ for a.a. } t \in [0, 1].$$

Theorem 4.7. *Let conditions* $(Q1)'$–$(Q2)'$ *and* $(Q3)$ *hold. Assume that there exists* $N > 0$ *such that for every* $u \in C(I, H)$, $\|u\|_2 > N$, *the following relation holds:*

$$\langle u, f \rangle_{L^2} > 0, \quad \text{for all } f \in Q(u). \tag{4.21}$$

Then problem (4.20) has a solution.

Proof. For each $n \in \mathbf{N}$, consider the auxiliary problem

$$\begin{cases} u'' \in P_n Q(u), \\ u(0) = u(1) = 0. \end{cases}$$

It is clear that this problem is equivalent to the following fixed point problem

$$u \in \Sigma_n(u), \tag{4.22}$$

where $\Sigma_n: C(I, H_n) \rightarrow K(C(I, H_n))$, $\Sigma_n(u) = J_n \circ Q(u)$.

From $(Q1)'$–$(Q2)'$ if follows that Σ_n is a completely u.s.c. CJ-multimap. Assume that $u \in C(I, H_n)$ is a solution of inclusion (4.22). Then there is $f \in Q(u)$ such that

$$u''(t) = P_n f(t) \quad \text{for a.a.} \quad t \in I.$$

We have

$$\langle u, f \rangle_{L^2} = \langle u, P_n f \rangle_{L^2} = \langle u, u'' \rangle_{L^2} \leq 0.$$

Therefore, $\|u\|_2 \leq N$. From $(Q2)'$ and (4.19) it follows that

$$\|u\|_C \leq q(1 + N^p).$$

Now set $R = qN^p + q + 1$ and define a multimap

$$\Psi_n \colon B_C^{(n)}(0, R) \times [0, 1] \to K(C(I, H_n)),$$
$$\Psi_n(u, \lambda) = J_n \circ \big(\delta(1 - \lambda)u + \lambda Q(u)\big),$$

where $0 < \delta < \frac{1}{N}$ and $B_C^{(n)}(0, R) = B_C(0, R) \cap C(I, H_n)$.

Following the method given in the proof of Theorem 4.1, we can obtain that Ψ_n is a compact *CJ*-multimap which has no fixed points on $\partial B_C^{(n)}(0, R) \times [0, 1]$. Then for sufficiently small δ we have

$$deg(i - J_n \circ Q, B_C^{(n)}(0, R)) = deg(i - \delta J_n \circ i, B_C^{(n)}(0, R))$$
$$= deg(i, B_C^{(n)}(0, R)) = 1.$$

Therefore, there is $u_n \in B_C(0, R) \cap W_0^{2,2}(I, H_n)$ such that $u_n'' = \mathbf{P}_n Q(u_n)$. $(Q3)$ implies that there exists $u_* \in B_C(0, R) \cap W_0^{2,2}(I, H)$ such that

$$u_*'' \in Q(u_*).$$

The function u_* is a solution of problem (4.20).   □

4.3.1   Application to a Second-Order Feedback Control System in Hilbert Space

Setting $Y = C[0, h]$ and $H = W^{1,2}[0, h]$ $(h > 0)$, we can consider the following feedback control system

$$
\begin{cases}
w''(t) = f(t, w(t), \varphi(t)), & \text{for a.a. } t \in [0, 1], \\
\varphi'(t) \in G(t, \varphi(t), w(t)), & \text{for a.a. } t \in [0, 1], \\
w(0) = w(1) = 0, \varphi(0) = 0,
\end{cases}
\tag{4.23}
$$

where $f \colon I \times Y \times Y \to Y$ is a upper Carathéodory map and $G \colon I \times Y \times Y \to Cv(H)$ is a u.s.c. multimap.

Assume that the map f and the multimap G satisfy the following conditions:

$(f1)'$ The restriction $f_{|I \times H \times Y}$ takes values in H.
$(f2)'$ There is $c > 0$ such that

$$\|f(t, y, z)\|_H \le c(1 + \|y\|_H + \|z\|_Y),$$

for all $(y, z) \in H \times Y$ and a.a. $t \in I$.

(G1)′ The multimap G is uniformly continuous with respect to the third argument in the following sense: for every $\varepsilon > 0$ there is $\delta > 0$ such that

$$G(t, y, \bar{z}) \subset O_\varepsilon(G(t, y, z)), \quad \forall (t, y) \in I \times Y$$

if $\|\bar{z} - z\|_Y < \delta$.

(G2)′ There is $d > 0$ such that

$$\|G(t, y, z)\|_H \le d(1 + \|y\|_Y + \|z\|_Y)$$

for all $(t, y, z) \in I \times Y \times Y$.

By a solution to problem (4.23) we mean a function $w \in W_0^{2,2}(I, H)$ such that there exists $\varphi \in W^{1,2}(I, H)$ with

$$\begin{cases} \varphi'(t) \in G(t, \varphi(t), w(t)), \text{ for a.a. } t \in I, \\ \varphi(0) = 0, \end{cases}$$

and

$$w''(t) = f(t, w(t), \varphi(t)), \text{ for a.a. } t \in I.$$

We need the following result.

Lemma 4.1 (see, Theorem 70.12 [64]). *Let E be a separable Banach space and $\Phi: I \times E \to Kv(E)$ be a multimap satisfying the following conditions:*

(Φ1) *for each $y \in E$ the multifunction $\Phi(\cdot, y)$ has a measurable selection;*
(Φ2) *for every $t \in I$ the multimap $\Phi(t, \cdot)$ is completely upper semicontinuous;*
(Φ3) *the set $\Phi(A)$ is compact for every compact subset $A \subset I \times E$;*
(Φ4) *there is $\omega \in L_+^2[0, 1]$ such that*

$$\|\Phi(t, y)\|_E \le \omega(t)(1 + \|y\|_E),$$

for all $(t, y) \in I \times E$.

Then the set of all solutions to the following problem

$$\begin{cases} g'(t) \in \Phi(t, g(t)), \ t \in I, \\ g(0) - g_0 \subset E, \end{cases}$$

is an R_δ-set in $C(I, E)$.

Theorem 4.8. *Let conditions $(f1)'$–$(f2)'$ and $(G1)'$–$(G2)'$ hold. Then problem (4.23) can be represented as problem (4.20) with conditions $(Q1)'$–$(Q2)'$ and $(Q3)$.*

Proof. Let us mention that Y is a separable space and the embedding $H \hookrightarrow Y$ is compact. From $(G2)'$ it follows that for every $(t, y, z) \in I \times Y \times Y$ the set $G(t, y, z)$ is bounded in H, hence it is a compact set in Y.

For a given function $w \in C(I, H)$ consider the following multimap

$$G_w: I \times Y \to Kv(Y), \quad G_w(t, y) = G(t, y, w(t)).$$

It is easy to verify that G_w satisfies conditions $(\Phi 1)$–$(\Phi 4)$. Notice that $(\Phi 4)$ follows from $(G2)'$ and the fact that for every $y \in H$ the following relation holds:

$$\|y\|_Y \le \max\{\sqrt{h}, \frac{1}{\sqrt{h}}\} \|y\|_H.$$

So we obtain that for every $w \in C(I, H)$ the set Ψ_w of all solutions to the following problem

$$\begin{cases} \varphi'(t) \in G(t, \varphi(t), w(t)), & t \in I \\ \varphi(0) = 0 \end{cases}$$

is an R_δ-set in $C(I, Y)$.

Define the multimap

$$\Psi: C(I, H) \to K(C(I, Y)), \quad \Psi(w) = \Psi_w.$$

From Theorem 5.2.5 [80] it follows that the multimap Ψ is upper semicontinuous.

Now set $\tilde{\Psi}: C(I, H) \to K\big(C(I, H) \times C(I, Y)\big)$

$$\tilde{\Psi}(w) = \{w\} \times \Psi(w),$$

and $\tilde{f}: C(I, H) \times C(I, Y) \to L^2(I, H)$,

$$\tilde{f}(w, \varphi)(t) = f(t, w(t), \varphi(t)).$$

Then problem (4.23) can be written in the form

$$\begin{cases} w'' \in Q(w), \\ w(0) = w(1) = 0, \end{cases}$$

where $Q: C(I, H) \to K(L^2(I, H)), \ Q(w) = \tilde{f} \circ \tilde{\Psi}(w)$.

We show now that the multimap Q satisfies conditions $(Q1)'$–$(Q2)'$ and $(Q3)$.

In fact, from the continuity of $\tilde{f}$ and the fact that

$$\tilde{\Psi} \in J\Big(C(I, H); C(I, H) \times C(I, Y)\Big)$$

it follows that $Q \in CJ\big(C(I, H); L^2(I, H)\big)$. So for every $n \in \mathbf{N}$ the restriction

$$(J_n \circ Q)|_{C(I,H_n)} \in CJ\big(C(I, H_n); C(I, H_n)\big).$$

Hence, condition $(Q1)'$ holds. Notice that condition $(Q2)'$ immediately follows from $(f2)'$, $(G2)'$ and Lemma 2.1.

Let us verify now the condition $(Q3)$. Let $M \subset W_0^{2,2}(I, H)$ be a bounded closed subset and assume that there are $\{n_k\}$ and $\{w_k\}$, $w_k \in M \cap W_0^{2,2}(I, H_{n_k})$, such that

$$w_k'' \in \mathbf{P}_{n_k} Q(w_k).$$

The set $\{w_k\}_{k=1}^\infty$ is bounded, so it is weakly compact. W.l.o.g. assume that

$$w_k \overset{W^{2,2}}{\rightharpoonup} w_0 \in M.$$

Therefore, $w_k'' \overset{L^2}{\rightharpoonup} w_0''$ and $w_k(t) \overset{H}{\rightharpoonup} w_0(t)$ for every $t \in I$. From the compactness of the embedding $H \hookrightarrow Y$ it follows that

$$w_k(t) \overset{Y}{\to} w_0(t), \tag{4.24}$$

for every $t \in I$.

Set $h_k \in Q(w_k)$, such that

$$w_k'' = \mathbf{P}_{n_k} h_k.$$

From $(Q2)'$ it follows that the set $\{Q(w_k)\}_{k=1}^\infty$ is bounded, hence the set $\{h_k\}_{k=1}^\infty$ is bounded in $L^2(I, H)$, and therefore it is weakly compact. W.l.o.g. assume that

$$h_k \overset{L^2}{\rightharpoonup} h_0 \in L^2(I, H).$$

Following the proof of Theorem 3.2 we have that $\mathbf{P}_{n_k} h_k \overset{L^2}{\rightharpoonup} h_0$. On the other hand

$$\mathbf{P}_{n_k} h_k = w_k'' \overset{L^2}{\rightharpoonup} w_0''.$$

So we obtain that $w_0'' = f_0$, i.e., $h_k \overset{L^2}{\rightharpoonup} w_0''$.

From $h_k \in Q(w_k)$ it follows that there is a sequence $\{\varphi_k\}_{k=1}^\infty$, $\varphi_k \in \Psi(w_k)$, such that

$$h_k(t) = f(t, w_k(t), \varphi_k(t)), \quad \textit{for a.a. } t \in I. \tag{4.25}$$

Set $\hat{W}^{1,2}(I, H) = \{u \in W^{1,2}(I, H) : u(0) = 0\}$. It is clear that $\hat{W}^{1,2}(I, H)$ is a subspace of $W^{1,2}(I, H)$. The set $\{\varphi_k\}_{k=1}^\infty$ is bounded in $\hat{W}^{1,2}(I, H)$ and so it is weakly compact. W.l.o.g. assume that

$$\varphi_k \overset{W^{1,2}}{\rightharpoonup} \varphi_0 \in \hat{W}^{1,2}(I, H).$$

Therefore

$$\varphi_k' \overset{L^2}{\rightharpoonup} \varphi_0' \ \ and \ \ \varphi_k(t) \overset{Y}{\to} \varphi_0(t), \ for \ every \ \ t \in I. \tag{4.26}$$

From $\varphi_k \in \Psi(w_k)$ it follows that there is $\{g_k\}_{k=1}^\infty \subset L^2(I, H)$ such that

$$g_k(t) \in G(t, \varphi_k(t), w_k(t)) \ \ for \ a.a. \ \ t \in I,$$

and

$$\varphi_k'(t) = g_k(t) \ \ for \ a.e. \ \ t \in I.$$

So $g_k \overset{L^2}{\rightharpoonup} \varphi_0'$. By the Mazur's Lemma (see, e.g., [44]) there are sequences of convex combinations $\{\hat{g}_m\}$ and $\{\hat{h}_m\}$

$$\hat{g}_m = \sum_{k=m}^\infty \lambda_{mk} g_k, \ \lambda_{mk} \geq 0 \ and \ \sum_{k=m}^\infty \lambda_{mk} = 1,$$

$$\hat{h}_m = \sum_{k=m}^\infty \tilde{\lambda}_{mk} h_k, \ \tilde{\lambda}_{mk} \geq 0 \ and \ \sum_{k=m}^\infty \tilde{\lambda}_{mk} = 1,$$

which converge in $L^2(I, H)$ to φ_0' and w_0'', respectively. Applying Theorem 38 [126] we again can assume w.l.o.g. that

$$\hat{g}_m(t) \overset{H}{\to} \varphi_0'(t) \ \ and \ \ \hat{h}_m(t) \overset{H}{\to} w_0''(t) \tag{4.27}$$

for a.e. $t \in I$.

From (4.24), (4.26) it follows that for every $t \in I$ and $\varepsilon > 0$ there is $i_0 = i_0(\varepsilon, t)$ such that

$$G(t, \varphi_i(t), w_i(t)) \subset O_\varepsilon^H \Big(G(t, \varphi_0(t), w_0(t)) \Big), \ \ for \ all \ i \geq i_0.$$

Then $g_i(t) \in O_\varepsilon^H \Big(G(t, \varphi_0(t), w_0(y)) \Big)$ for all $i \geq i_0$, and hence, from the convexity of the set $O_\varepsilon^H \Big(G(t, \varphi_0(t), w_0(t)) \Big)$ we have

$$\hat{g}_m(t) \in O_\varepsilon^H \Big(G(t, \varphi_0(t), w_0(t)) \Big), \ \ for \ all \ m \geq i_0.$$

Thus, $\varphi_0'(t) \in G(t, \varphi_0(t), w_0(t))$ for a.e. $t \in I$, i.e., $\varphi_0 \in \Psi(w_0)$.

Now from (4.24), (4.26) we have

$$\lim_{k \to \infty} f(t, w_k(t), \varphi_k(t)) = f(t, w_0(t), \varphi_0(t))$$

for a.e. $t \in I$.

So for a.e. $t \in I$ and $\varepsilon > 0$ there is $\tau_0 = \tau_0(\varepsilon, t)$ such that

$$f(t, w_\tau(t), \varphi_\tau(t)) \in O_\varepsilon^Y \left(f(t, w_0(t), \varphi_0(t)) \right), \quad \text{for all } \tau \ge \tau_0.$$

From (4.25) and (4.27) we obtain

$$w_0''(t) = f(t, w_0(t), \varphi_0(t)) \quad \text{for a.e. } t \in I.$$

So condition $(Q3)$ holds. □

4.3.2 Example

Now let the map f in (4.23) has the following form:

$$f(t, w(t), \varphi(t)) = b + aw(t) + \hat{f}(t, w(t), \varphi(t)),$$

where $a > 0, b \in \mathbf{R}$ and $\hat{f} \colon I \times Y \times Y \to H$ is a upper Carathéodory map.

Theorem 4.9. *Let conditions $(G1)'–(G2)'$ hold. Assume that the map $\hat{f}$ satisfies conditions $(f2)'$ and*

$(\hat{f})$ $a > c(1 + dr^2 e^{rd})$, *where* $r = \max\{\sqrt{h}, \frac{1}{\sqrt{h}}\}$ *and* c, d *are constants from* $(f2)'$ *and* $(G2)'$, *respectively.*

Then problem (4.23) has a solution.

Proof. It is easy to see that the map f satisfies conditions $(f1)'–(f2)'$. And hence, from Theorem 4.8 it follows that the multimap Q satisfies conditions $(Q1)'–(Q2)'$ and $(Q3)$.

Now for every $w \in W_0^{2,2}(I, H)$, choose an arbitrary $\gamma \in Q(w)$. Then there is a function $\varphi \in \Psi(w)$ such that

$$\gamma = b + aw + f^*(w, \varphi),$$

where $f^*(w, \varphi)(t) = \hat{f}(t, w(t), \varphi(t))$ for $t \in I$.

From $\varphi \in \Psi(w)$ it follows that there exists $g \in L^2(I, H)$ such that

$$g(t) \in G(t, \varphi(t), w(t)) \quad \text{for a.e. } t \in I,$$

and

$$\varphi(t) = \int_0^t g(s)ds \quad t \in I.$$

Therefore

$$\|\varphi(t)\|_H \le \int_0^t \|g(s)\|_H ds \le d \int_0^t (1 + \|\varphi(s)\|_Y + \|w(s)\|_Y) \, ds$$

$$\le d + dr \int_0^1 \|w(s)\|_H ds + \int_0^t rd \|\varphi(s)\|_H ds.$$

Using Lemma 2.1 we obtain

$$\|\varphi(t)\|_H \le \left(d + dr \int_0^1 \|w(s)\|_H ds\right) e^{rd} \quad \text{for all } t \in I.$$

Notice that for each $s \in I$: $u = w(s)$ and $v = \hat{f}(s, w(s), \varphi(s))$ are elements of H. We have

$$\left\langle w(s), b + aw(s) + \hat{f}(s, w(s), \varphi(s)) \right\rangle_H = \langle u, b + au + v \rangle_H$$

$$= \int_0^h a\left(u^2(\tau) + u'^2(\tau)\right)d\tau + b \int_0^h u(\tau)d\tau$$

$$+ \int_0^h \left(u(\tau)v(\tau) + u'(\tau)v'(\tau)\right)d\tau$$

$$\ge a\|u\|_H^2 - b\sqrt{h}\|u\|_H - \|u\|_H\|v\|_H.$$

So we obtain the following estimate

$$\langle w, \gamma \rangle_{L^2} = \int_0^1 \left\langle w(s), b + aw(s) + \hat{f}(s, w(s), \varphi(s)) \right\rangle_H ds$$

$$\ge \int_0^1 \left(a\|w(s)\|_H^2 - |b|\sqrt{h}\,\|w(s)\|_H - \|\hat{f}(s, w(s), \varphi(s))\|_H\,\|w(s)\|_H\right) ds$$

$$\ge a\|w\|_2^2 - |b|\sqrt{h}\|w\|_2 - \int_0^1 \|w(s)\|_H\, c\left(1 + \|w(s)\|_H + \|\varphi(s)\|_Y\right) ds$$

$$\ge (a - c)\|w\|_2^2 - (|b|\sqrt{h} + c)\|w\|_2 - cr \int_0^1 \|w(s)\|_H \|\varphi(s)\|_H \, ds$$

$$\geq (a-c)\|w\|_2^2 - (|b|\sqrt{h}+c)\|w\|_2 - cr \int_0^1 \|w(s)\|_H \, ds \left(d + dr \int_0^1 \|w(s)\|_H ds \right) e^{rd}$$

$$\geq (a - c - cdr^2 e^{rd})\|w\|_2^2 - (|b|\sqrt{h} + c + cdre^{rd})\|w\|_2 > 0$$

provided

$$\|w\|_2 > \frac{|b|\sqrt{h} + c + cdre^{rd}}{a - c - cdr^2 e^{rd}}.$$

From Theorem 4.7 it follows that problem (4.23) has a solution.   □

Chapter 5
Nonlinear Fredholm Inclusions and Applications

The necessity of studying coincidence points of nonlinear Fredholm operators and nonlinear (compact and condensing) maps of various classes arises in the investigation of many problems in the theory of partial differential equations and optimal control theory. Arising from the classic work of K.D. Elworthy and A.J. Tromba [45], the investigations of many researchers were directed on the study of topological characteristics of pairs consisting of nonlinear Fredholm operators and their perturbations of various types (see, e.g., [23, 27, 28, 135–137] and the references therein). For multivalued perturbations of nonlinear Fredholm operators the oriented coincidence degree was suggested in [23] for the case of convex-valued maps. Here, based on the works [98, 117, 134], we decribe a general construction of an oriented coincidence index for nonlinear Fredholm operators of zero index and approximable nonconvex-valued maps of compact and condensing type. A non-oriented analogue of such index was described earlier in the work [116]. Other constructions of an oriented coincidence index for nonlinear Fredholm operators can be found in [16, 17, 54, 130, 132] and the references therein.

We present also an application of the method of guiding functions in order to evaluate the oriented coincidence index. As an application, we consider a feedback control system, consisting of a first order implicit differential equation and a differential inclusion.

5.1 Preliminaries

By the symbols E, E' we denote real Banach spaces. Everywhere, by Y we denote an open bounded set $U \subset E$ (case (i)) or $U_* \subset E \times [0, 1]$ (case (ii)). We recall some notions (see, e.g., [27]).

Definition 5.1. A C^1-map $f : Y \to E'$ is Fredholm of index $k \geq 0$ $\left(f \in \Phi_k C^1 (Y)\right)$ if for every $y \in Y$ the Frechet derivative $f'(y)$ is a linear

V. Obukhovskii et al., *Method of Guiding Functions in Problems of Nonlinear Analysis*, 131
Lecture Notes in Mathematics 2076, DOI 10.1007/978-3-642-37070-0_5,
© Springer-Verlag Berlin Heidelberg 2013

Fredholm map of index k, that is, $\dim Ker\, f'(y) < \infty$, $\dim Co\, ker\, f'(y) < \infty$ and

$$\dim Ker\, f'(y) - \dim Co\, ker\, f'(y) = k \,.$$

Definition 5.2. A map $f : \overline{Y} \to E'$ is *proper* if $f^{-1}(\mathcal{K})$ is compact for every compact set $\mathcal{K} \subset E'$.

We recall now the notion of oriented Fredholm structure on Y.

An atlas $\{(Y_i, \Psi_i)\}$ on Y is said to be *Fredholm* if, for each intersecting charts (Y_i, Ψ_i) and (Y_j, Ψ_j) and every $y \in Y_i \cap Y_j$ it is

$$\left(\Psi_j \circ \Psi_i^{-1}\right)'(\Psi_i(y)) \in CG(\tilde{E}) \,,$$

where $\tilde{E}$ is the corresponding model space, and $CG(\tilde{E})$ denotes the collection of all linear invertible operators in $\tilde{E}$ of the form $i + k$, where i is the identity map and k is a compact linear operator.

The set $CG(\tilde{E})$ is divided into two connected components. The component containing the identity map is denoted by $CG^+(\tilde{E})$.

Two Fredholm atlases are said to be equivalent if their union is still a Fredholm atlas. The class of equivalent atlases is called a *Fredholm structure*.

A Fredholm structure on U is associated to a $\Phi_0 C^1$-map $f : U \to E'$ if it admits an atlas $\{(Y_i, \Psi_i)\}$ with model space E' for which

$$\left(f \circ \Psi_i^{-1}\right)'(\Psi_i(y)) \in LC(E')$$

at each point $y \in U$, where $LC(E')$ denotes the collection of all linear operators in E' of the form: identity plus a compact map. Let us note that each $\Phi_0 C^1$-map $f : U \to E'$ generates a Fredholm structure on U associated to f.

A Fredholm atlas $\{(Y_i, \Psi_i)\}$ on Y is said to be oriented if for each intersecting charts (Y_i, Ψ_i) and (Y_j, Ψ_j) and every $y \in Y_i \cap Y_j$ it is true that

$$\left(\Psi_j \circ \Psi_i^{-1}\right)'(\Psi_i(y)) \in CG^+(E) \,.$$

Two oriented Fredholm atlases are called oriently equivalent if their union is an oriented Fredholm atlas on Y. The equivalence class with respect to this relation is said to be the *oriented Fredholm structure on Y*.

We need also the following result (see [27]).

Proposition 5.1. Let $f \in \Phi_k C^1(Y)$; $K \subset Y$ a compact set. Then there exist an open neighborhood $\mathcal{O}$, $K \subset \mathcal{O} \subset Y$ and a finite dimensional subspace $E'_n \subset E'$ such that

$$f^{-1}(E'_n) \cap \mathcal{O} = M^{n+k} \,,$$

where M^{n+k} is a $n + k$ dimensional manifold. Moreover, the restriction $f_{|\mathcal{O}}$ is transversal to E'_n, i.e. $f'(x) E + E'_n = E'$ for each $x \in \mathcal{O}$.

In the sequel we use the following important property of ε-approximations.

Proposition 5.2. *Let X, X', Z be metric spaces; $f : X \to X'$ a continuous map; $\Sigma : X \to K(Z)$ an u.s.c. multimap; $\varphi : Z \to X'$ a continuous map. Suppose that $X_1 \subseteq X$ is a compact subset such that*

$$Coin(f, \varphi \circ \Sigma) \cap X_1 = \emptyset \,,$$

*where $Coin(f, \varphi \circ \Sigma) = \{x \in X : f(x) \in \varphi \circ \Sigma(x)\}$ is the **coincidence points** set. If $\varepsilon > 0$ is sufficiently small and $\sigma_\varepsilon \in a(\Sigma, \varepsilon)$, then*

$$Coin(f, \varphi \circ \sigma_\varepsilon) \cap X_1 = \emptyset$$

Proof. Suppose, to the contrary, that there is a sequences $\{x_n\} \subset X_1$ and $\varepsilon_n \to 0$, $\varepsilon_n > 0$ such that

$$f(x_n) = \varphi \sigma_{\varepsilon_n}(x_n) \tag{5.1}$$

for a sequence $\sigma_{\varepsilon_n} \in a(\Sigma, \varepsilon_n)$.

From Proposition 1.19 (i) and (ii) we can deduce that, w.l.o.g. the maps $\varphi \sigma_{\varepsilon_n|X_1}$ form a sequence of δ_n-approximations of $\varphi \Sigma_{|X_1}$, with $\delta_n \to 0$ and hence

$$(x_n, \varphi \sigma_{\varepsilon_n}(x_n)) \in O_{\delta_n}\left(\Gamma_{\varphi \Sigma_{|X_1}}\right).$$

The graph of the u.s.c. multimap $\varphi \Sigma_{|X_1}$ is a compact set (see, e.g. [80], Theorem 1.1.7), hence we can assume, w.l.o.g. that

$$(x_n, \varphi \sigma_{\varepsilon_n}(x_n)) \to (x_0, y_0) \in \Gamma_{\varphi \Sigma_{|X_1}} \ if \ n \to \infty$$

i.e. $y_0 \in \varphi \Sigma(x_0)$. Passing to the limit in (5.1), we obtain that $f(x_0) = y_0 \in \varphi \Sigma(x_0)$, i.e. $x_0 \in Coin(f, \varphi \Sigma)$, giving the contradiction. □

5.2 Oriented Coincidence Index

In this section we present the construction and describe the main properties of the oriented coincidence index for finite-dimensional, compact and condensing triplets.

We start from the following notion

Definition 5.3. The map $f : \overline{Y} \to E'$, the multimap $G = (\varphi \circ \Sigma) \in CJ(\overline{Y}, E')$ and the space $\overline{Y}$ form a compact triplet $(f, G, \overline{Y})_C$ if the following conditions are satisfied:

(h1) f is a continuous proper map, $f_{|Y} \in \Phi_k C^1(Y)$ with $k = 0$ in case (i), $k = 1$ in case (ii), and the Fredholm structure on Y generated by f is oriented;

(h2) G is compact, i.e. $G(\overline{Y})$ is a relatively compact subset of E';

(h3) $Coin(f, G) \cap \partial Y = \emptyset$

Let us mention that from hypotheses (h1), (h2) it follows that the coincidence points set $Q = Coin\,(f, G)$ is compact.

5.2.1 The Case of a Finite Dimensional Triplet

Given a triplet $(f, G, \overline{Y})_C$, from Proposition 5.1 we know that there exist an open neighborhood $\mathcal{O} \subset Y$ of the set $Q = Coin\,(f, G)$ and an n-dimensional subspace $E'_n \subset E'$ such that $f^{-1}\left(E'_n\right) \cap \mathcal{O} = M$, a manifold which is n-dimensional in case (i) and $(n + 1)$-dimensional in case (ii).

Now, suppose that the multimap $G = \varphi \circ \Sigma$ is finite dimensional, i.e. that there exists a finite dimensional subspace $E'_m \subset E'$ such that $G\left(\overline{Y}\right) \subset E'_m$. We can assume, w.l.o.g. that $E'_m \subset E'_n$. Then clearly $Q \subset M$. Let us mention also that the orientation on Y induces the orientation on M.

A compact triplet $(f, G, \overline{Y})_C$ such that G is finite dimensional is denoted by $(f, G, \overline{Y})_{C_m}$ and is called finite dimensional.

Lemma 5.1. For $(f, G = (\varphi \circ \Sigma), \overline{Y})_{C_m}$, let O_κ be a κ-neighborhood of Q. Then, $\Sigma_{|\overline{O}_\kappa}$ is approximable provided $\kappa > 0$ is sufficiently small.

Proof. Consider an open bounded set N satisfying the following conditions:

a) $Q \subset N \subset \overline{N} \subset M$;
b) $\overline{N}$ is a compact ANR-space.

Let us note that as N we can take an union of a finite collection of balls with centers in Q.
Let us take $\kappa > 0$ so that $O_\kappa \subset N$. Then the statement follows from Proposition 1.26 and Proposition 1.19 (i). $\square$

Now, let the neighborhood O_κ be chosen so that Σ is approximable on $\overline{O}_\kappa$. From Proposition 5.2 we know that

$$Coin\,(f, \varphi \circ \sigma_\varepsilon) \cap \partial O_\kappa = \emptyset$$

provided that $\sigma_\varepsilon \in a\left(\Sigma_{|\overline{O}_\kappa}, \varepsilon\right)$ and $\varepsilon > 0$ is sufficiently small.
So, we can consider the following map of pairs of spaces:

$$f - \varphi \circ \sigma_\varepsilon : \left(\overline{O}_\kappa, \partial O_\kappa\right) \to \left(E'_n, E'_n \backslash 0\right).$$

Now we are in position to give the following notion.

Definition 5.4. The oriented coincidence index of a finite dimensional triplet

$$\left(f, G = (\varphi \circ \Sigma), \overline{U}\right)_{C_m}$$

is defined by the equality

$$\left(f, G = (\varphi \circ \Sigma), \overline{U}\right)_{C_m} := \deg\left(f - \varphi \circ \sigma_\varepsilon, \overline{O}_\kappa\right) \tag{5.2}$$

where $\kappa > 0$ and $\varepsilon > 0$ are taken small enough and the right hand part of equality (5.2) denotes the Brouwer topological degree.

We want to prove that the given definition is consistent, i.e. the coincidence index does not depend on the choice of an ε-approximation σ_ε and the neighborhood O_κ.

Lemma 5.2. *Let σ_ε and $\sigma'_\varepsilon \in a\left(\Sigma_{|\overline{O}_\kappa}, \varepsilon\right)$ be two approximations. Then*

$$\deg\left(f - \varphi \circ \sigma_\varepsilon, \overline{O}_\kappa\right) = \deg\left(f - \varphi \circ \sigma'_\varepsilon, \overline{O}_\kappa\right) \tag{5.3}$$

if $\varepsilon > 0$ is sufficiently small.

Proof. Let us take any neighborhood N' of Q such that $Q \subset N' \subset \overline{N}' \subset O_\kappa$ and $\overline{N}'$ is an ANR-space. Then, by Proposition 1.19 (i) and Proposition 5.2 we know that we can take $\varepsilon > 0$ small enough so that $\sigma_{\varepsilon|\overline{N}'}$ and $\sigma'_{\varepsilon|\overline{N}'}$ are δ_0-approximations of $\Sigma_{|\overline{N}'}$ and

$$Coin\left(f, \varphi \circ \sigma_\varepsilon\right) \cap \left(\overline{O}_\kappa \backslash N'\right) = \emptyset \tag{5.4}$$

$$Coin\left(f, \varphi \circ \sigma'_\varepsilon\right) \cap \left(\overline{O}_\kappa \backslash N'\right) = \emptyset. \tag{5.5}$$

Since $\Sigma_{|\overline{N}'}$ is approximable, we can assume that $\varepsilon > 0$ is chosen so small that there exists a map $\gamma : \overline{N}' \times [0, 1] \to Z$ with the properties:

i) $\gamma\left(\cdot, 0\right) = \sigma_{\varepsilon|\overline{N}'}$, $\gamma\left(\cdot, 1\right) = \sigma'_{\varepsilon|\overline{N}'}$;

ii) $\gamma\left(\lambda, \cdot\right) \in a\left(\Sigma_{|\overline{N}'}, \delta_1\right)$ for each $\lambda \in [0, 1]$, where δ_1 is arbitrary small;

iii) $Coin\left(f, \varphi \circ \gamma\left(\cdot, \lambda\right)\right) \cap \partial N' = \emptyset$ for all $\lambda \in [0, 1]$. (see Propositions 1.26 and 5.2).

Each map $f - \varphi \circ \gamma\left(\cdot, \lambda\right)$, $\lambda \in [0, 1]$ transforms the pair $\left(\overline{N}', \partial N'\right)$ into the pair $(E_n, E_n \backslash 0)$ for each $\lambda \in [0, 1]$ and, by the homotopy property of the Brouwer degree we have that $\deg\left(f - \varphi \circ \sigma_\varepsilon, \overline{N}'\right) = \deg\left(f - \varphi \circ \sigma'_\varepsilon, \overline{N}'\right)$. Further from (5.4) and (5.5) and the additive property of the Brouwer degree we have

$$\deg\left(f - \varphi \circ \sigma_\varepsilon, \overline{O}_\kappa\right) = \deg\left(f - \varphi \circ \sigma_\varepsilon, \overline{N}'\right)$$

$$\deg\left(f - \varphi \circ \sigma'_\varepsilon, \overline{O}_\kappa\right) = \deg\left(f - \varphi \circ \sigma'_\varepsilon, \overline{N}'\right)$$

proving equality (5.3). $\qquad\qquad\qquad\qquad\qquad\qquad\qquad\qquad\qquad\qquad\qquad\square$

Now, if $O_{\kappa'} \subset O_{\kappa}$, the equality

$$\deg\left(f - \varphi \circ \sigma_{\varepsilon}, \overline{O}_{\kappa'}\right) = \deg\left(f - \varphi \circ \sigma_{\varepsilon}, \overline{O}_{\kappa}\right)$$

where $\varepsilon > 0$ is sufficiently small, follows easily from Propositions 1.19 (i), 5.2 and the additive property of the Brouwer degree.

At last, let us mention also the independence of the construction on the choice of the transversal subspace E_n'. In fact, if we take two subspaces E_{n_0}' and E_{n_1}', we can assume, w.l.o.g., that $E_{n_0}' \subset E_{n_1}'$. As earlier, we assume that $G\left(\overline{U}\right) \subset E_m' \subset E_{n_0}' \subset E_{n_1}'$. Then, from the construction we obtain two manifolds M^{n_0}, M^{n_1}, $M^{n_0} \subset M^{n_1}$ and two neighborhoods $O_{\kappa}^{n_0} \subset M^{n_0}$, $O_{\kappa}^{n_1} \subset M^{n_1}$, $O_{\kappa}^{n_0} \subset O_{\kappa}^{n_1}$ for $\kappa > 0$ sufficiently small. Now, take $\varepsilon > 0$ small enough, so that the degrees $\deg\left(f - \varphi \circ \sigma_{\varepsilon}, \overline{O_{\kappa}^{n_1}}\right)$ and $\deg\left(f - \varphi \circ \sigma_{\varepsilon}, \overline{O_{\kappa}^{n_0}}\right)$ are well defined. Then the equality

$$\deg\left(f - \varphi \circ \sigma_{\varepsilon}, \overline{O_{\kappa}^{n_1}}\right) = \deg\left(f - \varphi \circ \sigma_{\varepsilon}, \overline{O_{\kappa}^{n_0}}\right)$$

follows from the map restriction property of Brouwer degree.

Now, let us mention the main properties of the defined characteristic. Directly from Definition 5.4 and Proposition 5.2 we deduce the following statement.

Theorem 5.1 (The coincidence point property). *If* $Ind\left(f, G, \overline{U}\right)_{C_m} \neq 0$, *then*

$$\emptyset \neq Coin\left(f, G\right) \subset U.$$

To formulate the topological invariance property of the coincidence index, we give the following definition.

Definition 5.5. Two finite dimensional triplets

$$\left(f_0, G_0 = (\varphi_0 \circ \Sigma_0), \overline{U}_0\right)_{C_m} \quad \text{and} \quad \left(f_1, G_1 = (\varphi_1 \circ \Sigma_1), \overline{U}_1\right)_{C_m}$$

are said to be homotopic

$$\left(f_0, G_0 = (\varphi_0 \circ \Sigma_0), \overline{U}_0\right)_{C_m} \sim \left(f_1, G_1 = (\varphi_1 \circ \Sigma_1), \overline{U}_1\right)_{C_m}$$

if there exists a finite dimensional triplet $\left(f_*, G_*, \overline{U}_*\right)_{C_m}$, where $U_* \subset E \times [0, 1]$ is an open set, such that:

a) $U_i = U_* \cap (E \times \{i\})$, $i = 0, 1$;
b) $f_{*|\overline{U}_i} = f_i$, $i = 0, 1$;
c) G_* has the form

$$G_*\left(x, \lambda\right) = \varphi_*\left(\Sigma_*\left(x, \lambda\right), \lambda\right)$$

where $\Sigma_* \in J\left(\overline{U_*}, Z\right)$, $\varphi_* : Z \times [0, 1] \to E'$ is a continuous map, and

$$\Sigma_{*|\overline{U_i}} = \Sigma_i , \quad \varphi_{*|Z \times \{i\}} = \varphi_i , \quad i = 0, 1 .$$

Theorem 5.2 (The homotopy invariance property). *If*

$$\left(f_0, G_0, \overline{U}_0\right)_{C_m} \sim \left(f_1, G_1, \overline{U}_1\right)_{C_m} ,$$

then

$$\left| Ind\left(f_0, G_0, \overline{U}_0\right)_{C_m} \right| = \left| Ind\left(f_1, G_1, \overline{U}_1\right)_{C_m} \right| .$$

Proof. Let $\left(f_*, G_*, \overline{U_*}\right)_{C_m}$ be a finite dimensional triplet connecting the triplets $\left(f_0, G_0, \overline{U}_0\right)_{C_m}$ and $\left(f_1, G_1, \overline{U}_1\right)_{C_m}$. Let $O_{*\kappa} \subset U_*$ be a κ-neighborhood of $Q_* = Coin\left(f_*, G_*\right)$ where $\kappa > 0$ is sufficiently small.

Take $\sigma_{*\varepsilon} \in a\left(\Sigma_{*|\overline{O_{*\kappa}}}, \varepsilon\right)$ for $\varepsilon > 0$ sufficiently small. Applying Propositions 1.19 and 5.2 we can verify that the map $\varphi_* \circ \sigma_{*\varepsilon} : \overline{O_{*\kappa}} \to E'$,

$$\varphi_* \circ \sigma_{*\varepsilon}(x, \lambda) = \varphi_*\left(\sigma_{*\varepsilon}(x, \lambda), \lambda\right)$$

is a δ'-approximation of $G_{*|\overline{O_{*\kappa}}}$ for $\delta' > 0$ arbitrary small and, moreover

$$Coin\left(f_*, \varphi_* \circ \sigma_{*\varepsilon}\right) \cap \partial O_{*\kappa} = \emptyset$$

and $\varphi_* \circ \sigma_{*\varepsilon|\overline{O_{\kappa i}}}$, for $\overline{O_{\kappa i}} = \overline{O_{*\kappa}} \cap U_i$, $i = 0, 1$ are δ''-approximations of $G_{i|\overline{O_{\kappa i}}}$, $i = 0, 1$, where $\delta'' > 0$ is arbitrary small.
Denoting $\sigma_{*\varepsilon|\overline{O_{\kappa i}}} = \sigma_i$, $i = 0, 1$ we have that

$$\left| deg\left(f_0 - \varphi_0 \circ \sigma_0, \overline{O_{\kappa 0}}\right) \right| = \left| deg\left(f_1 - \varphi_1 \circ \sigma_1, \overline{O_{\kappa 1}}\right) \right|$$

(see [137]), proving the theorem.   □

Remark 5.1. If the Fredholm map f and the set U are constant under the homotopy, i.e. U_* has the form $U_* = U \times [0, 1]$ where $U \subset E$ is an open set and $f_*(x, \lambda) = f(x)$ for all $\lambda \in [0, 1]$, where $f \in \Phi_0 C^1(U)$, then

$$deg\left(f - \varphi_0 \circ \sigma_0, \overline{U}\right) = deg\left(f - \varphi_1 \circ \sigma_1, \overline{U}\right)$$

(see [136, 137]). Hence

$$Ind\left(f, G_0, \overline{U}\right)_{C_m} = Ind\left(f, G_1, \overline{U}\right)_{C_m} .$$

From Definition 5.4 and the additive property of the Brouwer degree we obtain the following property of the oriented coincidence index.

Theorem 5.3 (Additive dependence on the domain property). *Let U_0 and U_1 be disjoint open subsets of an open bounded set $U \subset E$ and $\left(f, G, \overline{U}\right)_{C_m}$ be a finite dimensional triplet such that*

$$Coin\left(f, G\right) \cap \left(\overline{U} \setminus \left(U_0 \cup U_1\right)\right) = \emptyset .$$

Then

$$Ind\left(f, G, \overline{U}\right)_{C_m} = \left(f, G, \overline{U_0}\right)_{C_m} + \left(f, G, \overline{U_1}\right)_{C_m}$$

5.2.2   The Case of a Compact Triplet

Now, we want to define the oriented coincidence degree for the general case of a compact triplet $\left(f, G = \left(\varphi \circ \Sigma\right), \overline{U}\right)_C$.

From the properness property of f and the compactness of G one can easily deduce the following statement.

Proposition 5.3. *Let $\left(f, G, \overline{U}\right)_C$ be a compact triplet; $\Lambda : \overline{U} \rightarrow K\left(E'\right)$ a multimap defined as*

$$\Lambda\left(y\right) = f\left(y\right) - G\left(y\right) .$$

Then, for every closed subset $U_1 \subset \overline{U}$, the set $\Lambda\left(U_1\right)$ is closed.

From the above assert it follows that, given a compact triplet $\left(f, G, \overline{U}\right)_C$, there exists $\delta > 0$ such that

$$B_\delta\left(0\right) \cap \Lambda\left(\partial U\right) = \emptyset \tag{5.6}$$

where $B_\delta\left(0\right) \subset E'$ is a δ-neighborhood of the origin.

Let us take a continuous map $i_\delta : G\left(\overline{U}\right) \rightarrow E_m$, where $E_m \subset E$ is a finite dimensional subspace, with the property that

$$\|i_\delta\left(v\right) - v\| < \delta \tag{5.7}$$

for each $v \in G\left(\overline{U}\right)$. As i_δ we can choose the Schauder projection (see, e.g. [95]).

Now, if G has the representation $G = \varphi \circ \Sigma$, consider the finite dimensional multimap $G_m = i_\delta \circ \varphi \circ \Sigma$. From (5.6) and (5.7) it follows that f, G_m and $\overline{U}$ form a finite dimensional triplet $\left(f, G_m, \overline{U}\right)_{C_m}$.

We can now define the oriented coincidence index for a compact triplet in the following way.

Definition 5.6. The oriented coincidence index of a compact triplet $\left(f, G = \left(\varphi \circ \Sigma\right), \overline{U}\right)_C$ is defined by the equality

$$Ind\left(f, G, \overline{U}\right)_C := Ind\left(f, G_m, \overline{U}\right)_{C_m}$$

where $G_m = i_\delta \circ \varphi \circ \Sigma$ and the map i_δ satisfies condition (5.7).

To prove the consistency of the given definition, it is sufficient to mention that, given two different maps $i_\delta^0, i_\delta^1 : \overline{G(U)} \to E_m'$ satisfying property (5.7), we have the homotopy of the corresponding finite-dimensional triplets:

$$\left(f, G_m^0, \overline{U}\right)_{C_m} \sim \left(f, G_m^1, \overline{U}\right)_{C_m}$$

where $G_m^k = i_\delta^k \circ \varphi \circ \Sigma$, $i = 0, 1$ (It is clear that the finite dimensional space E_m' can be taken the same for both maps i_δ^0, i_δ^1).

In fact, the homotopy is realized by the multimap $G_* : \overline{U} \times [0, 1] \to K\left(E_m'\right)$, defined as

$$G_*\left(x, \lambda\right) = \varphi_*\left(\Sigma\left(x, \lambda\right)\right) \text{ where } \varphi_*\left(z, \lambda\right) = (1 - \lambda)\, i_\delta^0 \varphi\left(z\right) + \lambda i_\delta^1 \varphi\left(z\right) .$$

So, from Remark 5.1 it follows that

$$Ind\left(f, G_m^0, \overline{U}\right)_{C_m} = Ind\left(f, G_m^1, \overline{U}\right)_{C_m} .$$

Applying Proposition 5.3 and Theorem 5.1 we can deduce the following coincidence point property

Theorem 5.4. *If $Ind\left(f, G, \overline{U}\right)_C \neq 0$ then $\emptyset \neq Coin\left(f, G\right) \subset U$.*

The definition of homotopy for compact triplets $\left(f_0, G_0, \overline{U}_0\right)_C \sim \left(f_1, G_1, \overline{U}_1\right)_C$ has the same form as in Definition 5.5 with the only difference that the connected triplet $\left(f_*, G_*, \overline{U}_*\right)$ is assumed to be compact.

Taking a finite dimensional approximation of $G_* = \varphi_* \circ \Sigma_*$ as $G_{*m} = i_\delta \circ \varphi_* \circ \Sigma_*$ and applying Theorem 5.2 and Definition 5.6, we obtain the following homotopy invariance property.

Theorem 5.5. *If $\left(f_0, G_0, \overline{U}_0\right)_C \sim \left(f_1, G_1, \overline{U}_1\right)_C$ then*

$$\left|Ind\left(f_0, G_0, \overline{U}_0\right)_C\right| = \left|Ind\left(f_1, G_1, \overline{U}_1\right)_C\right| .$$

Again, if f and U are constant, we have the equality

$$Ind\left(f, G_0, \overline{U}\right)_C = Ind\left(f, G_1, \overline{U}\right)_C .$$

An analogue of the additive dependence on the domain property (see Theorem 5.3) for compact triplets also holds.

5.2.3 Oriented Coincidence Index for Condensing Triplets

In this section we extend the notion of the oriented coincidence index to the case of condensing triplets. At first we recall some notions (see, e.g., [80]). Denote by

$P(E')$ the collection of all nonempty subsets of a Banach space E'. Let $(\mathscr{A}, \geq)$ be a partially ordered set.

Definition 5.7. A map $\beta : P(E') \to \mathscr{A}$ is called a measure of noncompactness (MNC) in E' if

$$\beta(\overline{co}\, D) = \beta(D) \quad \text{for every } D \in P(E').$$

A MNC β is called:

(i) monotone, if D_0, $D_1 \in P(E')$, $D_0 \subseteq D_1$ implies $\beta(D_0) \leq \beta(D_1)$;
(ii) nonsingular, if $\beta(\{a\} \cup D) = \beta(D)$ for every $a \in E'$, $D \in P(E')$;
(iii) real, if $A = \overline{\mathbf{R}_+} = [0, +\infty]$ with the natural ordering, and $\beta(D) < +\infty$ for every bounded set $D \in P(E')$.

Among the known examples of MNC satisfying all the above properties we can consider the *Hausdorff MNC*

$$\chi(D) = \inf\{\varepsilon > 0 : D \text{ has a finite } \varepsilon\text{--net}\} .$$

and the *Kuratowski MNC*

$$\alpha(D) = \inf\{d > 0 : D \text{ has a finite partition with sets of diameter less than } d\} .$$

Let again $Y = U \subset E$, or $U_* \subset E \times [0, 1]$, open bounded sets, $f : \overline{Y} \to E'$ a map; $G : \overline{Y} \to K(E')$ a multimap, β a MNC in E'.

Definition 5.8. Maps f, G and the space $\overline{Y}$ form a β-condensing triplet, $(f, G, \overline{Y})_\beta$ if they satisfy conditions (h1) and (h3) in Definition 5.3 and

h2$_\beta$) a multimap $G = \varphi \circ \Sigma \in CJ(\overline{Y}, E')$ is β-condensing w.r.t. f, i.e.

$$\beta(G(\Omega)) \not\geq \beta(f(\Omega))$$

for every $\Omega \subseteq \overline{Y}$ such that $G(\Omega)$ is not relatively compact.

Our target is to define the coincidence index for a β-condensing triplet $(f, G, \overline{U})_\beta$. To this aim, let us recall the following notion (see, e.g. [24, 57, 58, 80, 114]).

Definition 5.9. A convex, closed subset $T \subset E'$ is said to be fundamental for a triplet $(f, G, \overline{Y})_\beta$ if:

(i) $G(f^{-1}(T)) \subseteq T$;
(ii) for any point $y \in \overline{Y}$, the inclusion $f(y) \in \overline{co}(G(y) \cup T)$ implies that $f(y) \in T$.

The entire space E' and $\overline{co}G(\overline{Y})$ are natural examples of fundamental sets for $(f, G, \overline{U})_\beta$.

It is easy to verify the following properties of a fundamental set.

Proposition 5.4. *(a) The set Coin (f, G) is included in $f^{-1}(T)$ for each funda-*
mental set T of $(f, G, \overline{U})_\beta$;
(b) Let T be a fundamental set of $(f, G, \overline{U})_\beta$, and $P \subset T$, then the set $\tilde{T} =$
$\overline{co}\left(G\left(f^{-1}(T)\right) \cup P\right)$ is also fundamental;
(c) Let $\{T_\alpha\}$ be a system of fundamental sets of $(f, G, \overline{U})_\beta$. The set $T = \cap_\alpha T_\alpha$ is
also fundamental.

Proposition 5.5. *Each β-condensing triplet $(f, G, \overline{U})_\beta$, where β is a monotone,*
nonsingular MNC, admits a nonempty, compact fundamental set T.

Proof. Consider the collection $\{T_\alpha\}$ of all fundamental sets of $(f, G, \overline{U})_\beta$. containing an arbitrary point $a \in E'$. This collection is nonempty since it contains E'.
Then, taken $T = \cap_\alpha T_\alpha \neq \emptyset$ we obviously have

$$T = \overline{co}\left(G\left(f^{-1}(T)\right) \cup \{a\}\right)$$

and hence
$$\beta\left(f\left(f^{-1}(T)\right)\right) \leq \beta(T) = \beta\left(G\left(f^{-1}(T)\right)\right),$$

so $G\left(f^{-1}(T)\right)$ is relatively compact and T is compact.   □

Everywhere from now on, we assume that the MNC β is monotone and
nonsingular.

Now, if T is a nonempty compact fundamental set of a β-condensing triplet
$(f, G = (\varphi \circ \Sigma), \overline{Y})_\beta$, let $\rho : E' \to T$ be any retraction. Consider the multimap
$\tilde{G} = \rho \circ \varphi \circ \Sigma \in CJ\left(\overline{Y}, E'\right)$. From Proposition 5.4 (a) it follows that

$$Coin\left(f, \tilde{G}\right) = Coin\left(f, G\right) . \tag{5.8}$$

Hence, $f, \tilde{G}$, and $\overline{Y}$ form a compact triplet $(f, G, \overline{Y})_C$. We say that $(f, G, \overline{Y})_C$
is a compact approximation of the triplet $(f, G, \overline{Y})_\beta$.

Definition 5.10. The oriented coincidence index of a β-condensing triplet
$(f, G, \overline{U})_\beta$ is defined by the equality

$$Ind\left(f, G, \overline{U}\right)_\beta := Ind\left(f, \tilde{G}, \overline{U}\right)_C ,$$

where $(f, \tilde{G}, \overline{U})_C$ is a compact approximation of $(f, G, \overline{U})_\beta$.

To prove the consistency of the above definition, consider two nonempty,
compact fundamental sets T_0 and T_1 of the triplet $(f, G = \varphi \circ \Sigma, \overline{U})_\beta$ with
retractions $\rho_0 : E' \to T_0$ and $\rho_1 : E' \to T_1$ respectively.

If $T_0 \cap T_1 = \emptyset$, then by Proposition 5.4 (a), (c):

$$Coin\left(f, \tilde{G}_0\right) = Coin\left(f, \tilde{G}_1\right) = Coin\left(f, \tilde{G}\right) = \emptyset,$$

where $\tilde{G}_i = \rho_i \circ \varphi \circ \Sigma, i = 0, 1$. Hence, by Theorem 5.4:

$$Ind\ \left(f, \tilde{G}_0, \overline{U} \right)_C = Ind\ \left(f, \tilde{G}_1, \overline{U} \right)_C = 0.$$

Otherwise, we can assume, w.l.o.g., that $T_0 \subseteq T_1$. In this case, consider the map $\overline{\varphi} : Z \times [0, 1] \to E'$, given by $\overline{\varphi}(z, \lambda) = \rho_1 \circ (\lambda \varphi(z) + (1 - \lambda) \rho_0 \circ \varphi(z))$ and the multimap $\overline{G} \in CJ \left(\overline{U} \times [0, 1], E' \right), \overline{G}(x, \lambda) = \overline{\varphi}(\Sigma(x), \lambda)$.

The compact triplet $\left(f, \overline{G}, \overline{U} \times [0, 1] \right)_C$ realizes the homotopy

$$\left(f, \tilde{G}_0, \overline{U} \right)_C \sim \left(f, \tilde{G}_1, \overline{U} \right)_C.$$

Indeed, the only fact that we need to verify is that

$$Coin\ \left(\overline{f}, \overline{G} \right) \cap (\partial U \times [0, 1]) = \emptyset$$

where $\overline{f}(x, \lambda) \equiv f(x)$ is the natural extension.

To the contrary, suppose that there exists $(x, \lambda) \in \partial U \times [0, 1]$ such that

$$f(x) = \rho_1 \circ (\lambda \varphi(z) + (1 - \lambda) \rho_0 \circ \varphi(z))$$

for some $z \in \Sigma(x)$. But in this case, $x \in f^{-1}(T_1)$ and hence $\varphi(z) \in T_1$. Since also $\rho_0 \circ \varphi(z) \in T_1$ we have that

$$\lambda \varphi(z) + (1 - \lambda) \rho_0 \circ \varphi(z) \in T_1$$

and so

$$f(x) = \lambda \varphi(z) + (1 - \lambda) \rho_0 \circ \varphi(z) \in \overline{co}\,(G(x) \cup T_0)$$

and we obtain that $f(x) \in T_0$ and $x \in f^{-1}(T_0)$, implying $\varphi(z) \in T_0$ and $\rho_0 \circ \varphi(z) = \varphi(z)$. We conclude that $f(x) = \varphi(z) \in G(x)$ giving a contradiction.

Definition 5.11. Two β-condensing triplets $\left(f_0, G_0, \overline{U}_0 \right)_\beta$ and $\left(f_1, G_1, \overline{U}_1 \right)_\beta$ are said to be homotopic,

$$\left(f_0, G_0, \overline{U}_0 \right)_\beta \sim \left(f_1, G_1, \overline{U}_1 \right)_\beta$$

if there exists a β-condensing triplet $\left(f_*, G_*, \overline{U}_* \right)_\beta$ satisfying conditions (a), (b), (c) of Definition 5.5.

Theorem 5.6 (The homotopy invariance property). *If*

$$\left(f_0, G_0, \overline{U}_0 \right)_\beta \sim \left(f_1, G_1, \overline{U}_1 \right)_\beta,$$

then

$$\left| Ind\ \left(f_0, G_0, \overline{U}_0 \right)_\beta \right| = \left| Ind\ \left(f_1, G_1, \overline{U}_1 \right)_\beta \right|. \tag{5.9}$$

Proof. Let T_* be a nonempty compact fundamental set of the triplet

$$(f_*, G_* = (\varphi_* \circ \Sigma_*), \overline{U}_*)$$

connecting $(f_0, G_0, \overline{U}_0)_\beta$ with $(f_1, G_1, \overline{U}_1)_\beta$. It is easy to see that T_* is fundamental also for the triplets $(f_k, G_k, \overline{U}_k)_\beta, k = 0, 1.$
Let $\rho_* : E' \to T_*$ be any retraction, and $(f_*, \tilde{G}_* = \rho_* \circ \varphi_* \circ \sigma_*, \overline{U}_*)_C$ the corresponding compact approximation of $(f_*, G_*, \overline{U}_*)_\beta$. Then $(f_*, \tilde{G}_*, \overline{U}_*)_C$ realizes a compact homotopy connecting the triplets $(f_k, \rho_* \circ \varphi_k \circ \Sigma_k, \overline{U}_k)_C, k = 0, 1$ which are compact approximations of $(f_k, G_k, \overline{U}_k)_\beta, k = 0, 1$ respectively.
By Theorem 5.5 we have

$$\left| Ind\, (f_0, \rho_* \circ \varphi_0 \circ \Sigma_0, \overline{U}_0)_C \right| = \left| Ind\, (f_1, \rho_* \circ \varphi_1 \circ \Sigma_1, \overline{U}_1)_C \right|$$

giving the desired equality (5.9). □

Remark 5.2. Let us mention that in case of constant f and $\overline{U}$:

$$U_* = U \times [0, 1]$$
$$f_* (x, \lambda) \equiv f (x) , \quad \forall \lambda \in [0, 1] ,$$

the condition of β-condensivity for a triplet $(f, G_*, \overline{U} \times [0, 1])_\beta$ can be weakened: for the existence of a nonempty, compact fundamental set T it is sufficient to demand that

$$\beta (G_* (\Omega \times [0, 1])) \ngeq \beta (f (\Omega))$$

for every $\Omega \subseteq \overline{U}$ such that $G_* (\Omega \times [0, 1])$ is not relatively compact.
In fact, it is enough to notice that in this case $f_*^{-1} (T) = f^{-1} (T) \times [0, 1]$ and to follow the line of reasoning of Proposition 5.5.
Taking into consideration the corresponding property of compact triplets, we can precise the above property of homotopy invariance.
If $(f, G_*, \overline{U} \times [0, 1])_\beta$ is a β-condensing triplet, where G_* has the form (c) of Definition 5.5, then

$$Ind\, (f, G_0, \overline{U})_\beta = Ind\, (f, G_1, \overline{U})_\beta$$

where $G_k = G_* (\cdot, \{k\}), k = 0, 1.$

From relation (5.8) and Theorem 5.4, follows immediately

Theorem 5.7 (Coincidence Point Property). *If* $Ind\, (f, G, \overline{U})_\beta \neq 0$, *then*

$$\emptyset \neq Coin\, (f, G) \subset U.$$

As an example of application of Theorems 5.6 and 5.7 consider the following coincidence point result.

Theorem 5.8. *Let* $f \in \Phi_0 C^1(E, E')$ *be odd;* $G \in CJ(E, E')$ β-*condensing w.r.t.* f *on bounded subsets of* E, *i.e.* $\beta(G(\Omega)) \not\geq \beta(f(\Omega))$ *for every bounded set* $\Omega \subset E$ *such that* $G(\Omega)$ *is not relatively compact.*
If the set of solutions of one-parameter family of operator inclusions

$$f(x) \in \lambda G(x) \tag{5.10}$$

is a priori bounded, then $Coin(f, G) \neq \emptyset$.

Proof. From the condition it follows that there exists a ball $\mathscr{B} \subset E$ centered at the origin whose boundary $\partial \mathscr{B}$ does not contain solutions of (5.10).
Let $\varphi \circ \Sigma$ be a representation of G. If $G_* : \overline{\mathscr{B}} \times [0, 1] \to K(E')$ has the form

$$G_*(z, \lambda) = \varphi_*(\Sigma(z), \lambda) \quad \text{and}$$

$$\varphi_*(z, \lambda) = \lambda \varphi(z)$$

then f, G_* and $\overline{\mathscr{B}} \times [0, 1]$ form a β-condensing triplet $\left(f, G_*, \overline{\mathscr{B}} \times [0, 1]\right)_\beta$.

In fact, suppose that $\beta(G_*(\Omega)) \geq \beta(f(\Omega))$ for some $\Omega \subset \overline{\mathscr{B}}$. Since

$$G_*(\Omega \times [0, 1]) = \overline{co}(G(\Omega) \cup \{0\})$$

we have that $\beta(G(\Omega)) \geq \beta(f(\Omega))$ implying that $G(\Omega)$ and hence $G_*(\Omega \times [0, 1])$ is relatively compact.
So the triplet $\left(f, G_*, \overline{\mathscr{B}} \times [0, 1]\right)_\beta$ induces an homotopy connecting the triplets $\left(f, G, \overline{\mathscr{B}}\right)_\beta$ and $\left(f, 0, \overline{\mathscr{B}}\right)_\beta$. Since the triplet $\left(f, 0, \overline{\mathscr{B}}\right)_\beta$ is finite dimensional, from the odd condition on f and the odd field property of the Brouwer degree, it follows that $\left(f, 0, \overline{\mathscr{B}}\right)_\beta$ is an odd number.
Then, from the equality $Ind\left(f, G, \overline{\mathscr{B}}\right)_\beta = Ind\left(f, 0, \overline{\mathscr{B}}\right)_\beta$ it follows that

$$Ind\left(f, G, \overline{\mathscr{B}}\right)_\beta \neq 0$$

and we can apply the coincidence point property.   □

In conclusion of this section, let us formulate the additive dependence on the domain property for β-condensing triplets.

Theorem 5.9. *Let U_0 and U_1 be disjoint open subsets of an open bounded set $U \subset E$. If $\left(f, G, \overline{U}\right)_\beta$ is a β-condensing triplet such that*

$$Coin\,(f, G) \cap \left(\overline{U} \setminus (U_0 \cup U_1)\right) = \emptyset,$$

then,

$$Ind\,\left(f, G, \overline{U}\right)_\beta = Ind\,\left(f, G, \overline{U}_0\right)_\beta + Ind\,\left(f, G, \overline{U}_1\right)_\beta.$$

5.3 Calculation of the Oriented Coincidence Index by the MGF

In this section, the method of guiding functions is used to calculate the oriented coincidence index for a class of feedback control systems. This characteristic allows to obtain the existence result for periodic trajectories of such systems.

We consider the abstract results in Sect. 5.3.1, while an illustrating example is given in Sect. 5.3.2.

5.3.1 The Main Result

Consider a feedback control system of the following form:

$$A\left(t, x(t), x'(t)\right) = B\left(t, x(t), y(t)\right), \ for\ t \in [0, 1], \tag{5.11}$$

$$y'(t) \in C\left(t, x(t), y(t)\right), \ for\ a.a.\ t \in [0, 1], \tag{5.12}$$

$$x(0) = x(1), \tag{5.13}$$

$$y(0) = y_0, \tag{5.14}$$

where $A: [0, 1] \times \mathbf{R} \times \mathbf{R} \to \mathbf{R}$ and $B: [0, 1] \times \mathbf{R} \times \mathbf{R} \to \mathbf{R}$ are continuous maps; $C: [0, 1] \times \mathbf{R} \times \mathbf{R} \to Kv(\mathbf{R})$ is an upper Carathéodory multimap; $y_0 \in \mathbf{R}$. Here $x : [0, 1] \to \mathbf{R}$ is *the state function* and $y : [0, 1] \to \mathbf{R}$ is *the control function*.

Let us denote by $C^1[0, 1]$ [$AC[0, 1]$] the collection of all continuously differentiable [resp., absolutely continuous] functions on $[0, 1]$. Set

$$C_{pr}^1[0, 1] = \{x \in C^1[0, 1]: x(0) = x(1)\}.$$

The norms of elements $x \in C^1[0, 1]$ and $y \in C[0, 1]$ are denoted by $\|x\|_{C^1}$ and $\|y\|_C$, respectively. By symbol $B_{C_{pr}^1}(0, R)$ we denote the closed ball of radius R in $C_{pr}^1[0, 1]$.

By *a solution of problem* (5.11)–(5.14) we mean a couple of functions $x \in C^1_{pr}[0, 1]$ (*the trajectory*) and $y \in AC[0, 1]$ (*the control*), satisfying relations (5.11)–(5.12) and conditions (5.13)–(5.14).

We assume that the following conditions are fulfilled:

($A1$) There exist continuous partial derivatives $A'_u(t, u, v)$, $A'_v(t, u, v)$ and moreover, $A'_v(t, u, v) \neq 0$ for all (t, u, v).

($A2$) $A'_v(t, 0, 0) = 1$ for all $t \in [0, 1]$.

($A3$) There is a positive function $\alpha \in C[0, 1]$ such that for every $u, v, w \in \mathbf{R}$ satisfying

$$(1 - \lambda)v + \lambda A(t, u, v) = w \quad \textit{for any} \ \lambda \in [0, 1]$$

we have

$$|v| \leq \alpha(t)(1 + |w| + |u|)$$

for all $t \in [0, 1]$.

(B) There exists a constant $c > 0$ such that

$$|B(t, u, v)| \leq c(1 + |u| + |v|),$$

for all $(t, u, v) \in [0, 1] \times \mathbf{R} \times \mathbf{R}$.

($C1$) The multimap C is uniformly continuous in the second argument in the following sense: for each $\varepsilon > 0$ there exists $\delta > 0$ such that

$$C(t, \bar{u}, v) \subset O_\varepsilon\big(C(t, u, v)\big) \ \textit{for all} \ (t, v) \in [0, 1] \times \mathbf{R},$$

whenever $|\bar{u} - u| < \delta$ (here O_ε denotes the ε-neighborhood of a set).

($C2$) There exists a constant $d > 0$ such that

$$\|C(t, u, v)\| := \max\{|z| : z \in C(t, u, v)\} \leq d(1 + |u| + |v|),$$

for all $(u, v) \in \mathbf{R} \times \mathbf{R}$ and a.a. $t \in [0, 1]$.

Definition 5.12. A continuously differentiable function $V : \mathbf{R} \to \mathbf{R}$ is said to be an integral guiding function for problem (5.11)–(5.14), if there exists $N > 0$ such that for every $x \in C^1_{pr}[0, 1]$, $\|x\|_2 \geq N$, the following relations hold:

$$\int_0^1 A(t, x(t), x'(t))V'(x(t))\, dt \leq 0;$$

$$\int_0^1 B(t, x(t), y(t))V'(x(t))\, dt > 0$$

for all $y \in AC[0, 1]$ satisfying (5.12) and (5.14).

From this definition it follows that if V is a guiding function for problem (5.11)–(5.14), then $V'(a) \neq 0$ provided $|a| \geq r \geq N$. Then the topological degree $deg(V', [-r, r])$ is well defined and it does not depend on $r \geq N$. This number is denoted by $Ind\, V$.

Theorem 5.10. *Let conditions (A1)–(A3), (B) and (C1)–(C2) hold. In addition, assume that there is a guiding function V for problem (5.11)–(5.14) such that $Ind\, V \neq 0$. Then problem (5.11)–(5.14) has a solution.*

Proof. The proof is divided into the following steps.

STEP 1. Following the methods of [117], Proposition 5.1, we show that under condition (A1) the map $f : C_{pr}^1[0, 1] \to C[0, 1]$,

$$f(x)(t) = A(t, x(t), x'(t)), \tag{5.15}$$

is a Fredholm map of zero index, whose restriction to each closed bounded set $D \subset C_{pr}^1[0, 1]$ is proper.

In fact, let us note that f is a C^1 map and, moreover, its derivative can be written explicitly:

$$\left(f'(x)\, h \right)(t) = A_u' \left(t, x(t), x'(t) \right) h(t) + A_v' \left(t, x(t), x'(t) \right) h'(t)$$

for $h \in C_{pr}^1[0, 1]$. Introducing the auxiliary operators

$$f_u'(x) : C_{pr}^1[0, 1] \to C[0, 1],$$

$$\left(f_u'(x)\, h \right)(t) = A_u' \left(t, x(t), x'(t) \right) h(t), \quad t \in [0, 1]$$

and

$$f_v'(x) : C_{pr}^1[0, 1] \to C[0, 1],$$

$$\left(f_v'(x)\, h \right)(t) = A_v' \left(t, x(t), x'(t) \right) h'(t), \quad t \in [0, 1],$$

we can write

$$f'(x)\, h = f_u'(x)\, h + f_v'(x)\, h.$$

The operator $f_u'(x)$ is completely continuous since it can be represented as the composition of a completely continuous injection map $i : C_{pr}^1[0, 1] \to C[0, 1]$ and a continuous linear operator $M : C[0, 1] \to C[0, 1]$,

$$(Mh)(t) = A_u' \left(t, x(t), x'(t) \right) h(t).$$

Now, it is sufficient to show that the operator $f_v'(x)$ is a linear Fredholm operator of zero index.

Let us represent this operator as the composition of the differentiation operator $d/dt : C_{pr}^1[0, 1] \to C[0, 1]$ and the operator $J : C[0, 1] \to C[0, 1]$,

$$(Jz)(t) = A_v' \left(t, x(t), x'(t) \right) z(t).$$

It is well known that the operator d/dt is a linear Fredholm operator of index zero. Since the value $A'_v(t, x(t), x'(t))$ is non-zero, the operator J is invertible too. Hence, the operators $f'_v(x)$ and, therefore, $f'(x)$ are Fredholm of zero index. So, f is a nonlinear Fredholm map of zero index.

Now, let $D \subset C^1_{pr}[0, 1]$ be a closed bounded set. Denoting the restriction of f on D by the same symbol, let us demonstrate that it is proper. Let $\mathcal{K} \subset C[0, 1]$ be any compact set, and $\{x_n\}_{n \in \mathbb{N}} \subset f^{-1}(\mathcal{K})$ be an arbitrary sequence. W.l.o.g. we can assume that $f(x_n) \to z \in \mathcal{K}$. Since the sequence $\{x_n\}$ is bounded in $C^1_{pr}[0, 1]$ we can also assume, w.l.o.g., that the sequence $\{x_n\}$ tends, in $C[0, 1]$, to some $\omega \in C[0, 1]$. Further, from the representation

$$A\left(t, \omega(t), x'_n(t)\right) = A\left(t, x_n(t), x'_n(t)\right)$$

$$+ \left[A\left(t, \omega(t), x'_n(t)\right) - A\left(t, x_n(t), x'_n(t)\right)\right]$$

it follows that the sequence $z_n = A\left(\cdot, \omega(\cdot), x'_n(\cdot)\right)$ tends to z in $C[0, 1]$. From the inverse mapping theorem it follows that $x'_n = \Psi(z_n)$, where $\Psi : C[0, 1] \to C[0, 1]$ is a continuous map, implying that x'_n tends to $\Psi(z)$ in $C[0, 1]$. So, the sequence $\{x_n\}_{n \in \mathbb{N}}$ is convergent in the space $C^1_{pr}[0, 1]$ and, hence, the set $f^{-1}(\mathcal{K})$ is compact.

STEP 2. Now we show that the set of trajectories of system (5.11)–(5.14) is a priori bounded. To do this, for a given function $x \in C[0, 1]$ define the multimap

$$C_x : [0, 1] \times \mathbb{R} \to Kv(\mathbb{R}), \quad C_x(t, w) = C(t, x(t), w).$$

Theorem 1.3.5, [80] implies that, for each $w \in \mathbb{R}$, the multifunction $C_x(\cdot, w)$ has a measurable selection. Further, from condition $(C1)$ and the fact that C is an upper Carathéodory multimap it follows that for a.e. $t \in [0, 1]$ the multimap $C_x(t, w)$ upper semicontinuously depends on (w, x).

From Proposition 2.1 and the continuous dependence of the solution set of a differential inclusion on a parameter (see, e.g. [80]) we know that: for each $x \in C[0, 1]$ the set of solutions Π_x of the Cauchy problem

$$\begin{cases} y'(t) \in C(t, x(t), y(t)) \text{ for a.a. } t \in [0, 1] \\ y(0) = y_0 \end{cases}$$

is an R_δ-set in $C[0, 1]$ and the multimap

$$\Pi : C[0, 1] \to K(C[0, 1]), \quad \Pi(x) = \Pi_x,$$

is upper semicontinuous.

Now let the multimap $\tilde{\Pi} : C[0, 1] \to K(C[0, 1] \times C[0, 1])$ be defined as

$$\tilde{\Pi}(x) = \{x\} \times \Pi(x),$$

and the map $\tilde{B}: C[0, 1] \times C[0, 1] \to C[0, 1]$ is given by the formula

$$\tilde{B}(x, y)(t) = B(t, x(t), y(t)).$$

Define $\tilde{G}: C[0, 1] \to K(C[0, 1])$ as the composition

$$\tilde{G}(x) = \tilde{B} \circ \tilde{\Psi}(x),$$

and let G be the restriction of $\tilde{G}$ on $C_{pr}^1[0, 1]$. It is easy to see that G is a completely u.s.c. CJ-multimap and we can reduce problem (5.11)–(5.14) to the following operator inclusion

$$f(x) \in G(x). \tag{5.16}$$

Now let us assume that $x_* \in C_{pr}^1[0, 1]$ is a solution of inclusion (5.16). Then there is a function $y_* \in \Pi(x_*)$ such that

$$A(t, x_*(t), x_*'(t)) = B(t, x_*(t), y_*(t)), \text{ for all } t \in [0, 1]. \tag{5.17}$$

Therefore

$$\int_0^1 A(t, x_*(t), x_*'(t)) V'(x_*(t)) \, dt = \int_0^1 B(t, x_*(t), y_*(t)) V'(x_*(t)) \, dt,$$

and hence, $\|x_*\|_2 \le N$.

From $y_* \in \Pi(x_*)$ it follows that there is $g_* \in L^1[0, 1]$ such that

$$g_*(t) \in C(t, x_*(t), y_*(t)), \text{ for a.a. } t \in [0, 1], \text{ and}$$

$$y_*(t) = y_0 + \int_0^t g_*(s) \, ds.$$

By condition $(C2)$, for every $t \in [0, 1]$ the following estimation holds:

$$|y_*(t)| \le |y_0| + \int_0^t d(1 + |x_*(s)| + |y_*(s)|) \, ds$$

$$\le |y_0| + d + d\|x_*\|_2 + \int_0^t d|y_*(s)| \, ds.$$

Applying Lemma 2.1, we obtain that

$$|y_*(t)| \le (|y_0| + d + d\|x_*\|_2) e^d. \tag{5.18}$$

From (5.17)–(5.18), (B) and $(A3)$ we have

$$|x'_*(t)| \leq \alpha(t)\Big(1 + |x_*(t)| + |B\big(t, x_*(t), y_*(t)\big)|\Big)$$

$$\leq \alpha(t)\Big(1 + |x_*(t)| + c\big(1 + |x_*(t)| + |y_*(t)|\big)\Big)$$

$$\leq (c+1)\alpha(t) + (c+1)\alpha(t)|x_*(t)| + c\alpha(t)\big(|y_0| + d + d\|x_*\|_2\big)e^d.$$

$$(5.19)$$

Therefore

$$\|x'_*\|_2 \leq (c+1)\|\alpha\|_2 + (c+1)\|\alpha\|_2\|x_*\|_2 + c\|\alpha\|_2\big(|y_0| + d + d\|x_*\|_2\big)e^d$$

$$\leq (c+1)\|\alpha\|_2 + (c+1)N\|\alpha\|_2 + c\|\alpha\|_2\big(|y_0| + d + dN\big)e^d = M.$$

So, $\|x_*\|_C \leq \|x_*\|_2 + \|x'_*\|_2 \leq N + M$.
From (5.19) it follows that

$$\|x'_*\|_C \leq (c+1)\|\alpha\|_C + (c+1)(M+N)\|\alpha\|_C + c\|\alpha\|_C\big(|y_0| + d + dN\big)e^d = K.$$

Hence, $\|x_*\|_{C^1} \leq R = M + N + K$.

STEP 3. Now, choosing an arbitrary $R' > R$, we evaluate the coincidence index of the compact triplet $\big(f, G, B_{C^1_{pr}}(0, R')\big)$. To prove it, set

$$A^{\sharp} : [0,1] \times \mathbf{R} \times \mathbf{R} \times [0,1] \to \mathbf{R},$$

$$A^{\sharp}(t, u, v, \lambda) = v + \lambda\big(A(t, u, v) - v\big),$$

and define

$$f^{\sharp} : C^1_{pr}[0,1] \times [0,1] \to C[0,1],$$

$$f^{\sharp}(x, \lambda)(t) = A^{\sharp}\big(t, x(t), x'(t), \lambda\big). \qquad (5.20)$$

It is easy to see that there exist continuous partial derivatives $A^{\sharp'}_u(t, u, v, \lambda)$ and $A^{\sharp'}_v(t, u, v, \lambda)$.

Let us demonstrate that $A^{\sharp'}_v(t, u, v, \lambda) \neq 0$ for all (t, u, v, λ). Assume, to the contrary, that there is $(t_0, u_0, v_0, \lambda_0)$ such that

$$A^{\sharp'}_v(t_0, u_0, v_0, \lambda_0) = 0,$$

or equivalently, $1 + \lambda_0\big(A'_v(t_0, u_0, v_0) - 1\big) = 0$. It is clear that $\lambda_0 > 0$. From $(A1)$–$(A2)$ it follows that $A'_v(t, u, v) > 0$ for all $(t, u, v) \in [0,1] \times \mathbf{R} \times \mathbf{R}$. Therefore,

$$\lambda_0\big(A'_v(t_0, u_0, v_0) - 1\big) > -\lambda_0,$$

and hence, $1 + \lambda_0\big(A'_v(t_0, u_0, v_0) - 1\big) > 1 - \lambda_0 \geq 0$, giving a contradiction.

As in Step 1, we obtain that $f^\sharp$ is a Fredholm operator of index 1, whose restriction to $B_{C^1_{pr}}(0, R') \times [0, 1]$ is proper. Notice that the Fredholm structure on $B_{C^1_{pr}}(0, R') \times [0, 1]$ generated by $f^\sharp$ is oriented.

Consider now the triplet $\left(f^\sharp, G^\sharp, B_{C^1_{pr}}(0, R') \times [0, 1]\right)$, where

$$G^\sharp : B_{C^1_{pr}}(0, R') \times [0, 1] \to K\left(C[0, 1]\right), G^\sharp(x, \lambda) = G(x).$$

It is clear that $G^\sharp$ is a compact CJ-multimap.

Assume that there is $(x, \lambda) \in \partial B_{C^1_{pr}}(0, R') \times [0, 1]$ such that

$$f^\sharp(x, \lambda) \in G^\sharp(x, \lambda).$$

Then there exists $y \in \Pi(x)$ such that

$$A^\sharp(t, x(t), x'(t), \lambda) = B(t, x(t), y(t)) \ for \ all \ t \in [0, 1].$$

Therefore

$$\int_0^1 B(t, x(t), y(t)) V'(x(t))\, dt = \int_0^1 A^\sharp(t, x(t), x'(t), \lambda) V'(x(t))\, dt$$

$$= \lambda \int_0^1 A(t, x(t), x'(t)) V'(x(t))\, dt.$$

So, $\|x\|_2 \le N$, and hence, from (5.18), (B) and $(A3)$ we obtain that

$$\|x\|_{C^1} \le R < R',$$

giving a contradiction.

Thus, the compact triplet $\left(f^\sharp, G^\sharp, B_{C^1_{pr}}(0, R') \times [0, 1]\right)$ is a homotopy joining the compact triplets $\left(f, G, B_{C^1_{pr}}(0, R')\right)$ and $\left(L, G, B_{C^1_{pr}}(0, R')\right)$, where $Lx = x'$. Using Theorem 5.5 we obtain

$$\mid Ind\left(f, G, B_{C^1_{pr}}(0, R')\right) \mid = \mid Ind\left(L, G, B_{C^1_{pr}}(0, R')\right) \mid .$$

Further, L is a linear Fredholm operator of zero index and

$$Ker\, L \cong \mathbf{R} \cong Coker\, L.$$

The projection $\Pi_L : C[0, 1] \to \mathbf{R}$ is defined as

$$\Pi_L(w) = \int_0^1 w(s)\,ds,$$

and the homeomorphism $\Lambda_L \colon \mathbf{R} \to \mathbf{R}$ is an identity map. The space $C[0,1]$ can be represented as

$$C[0,1] = \mathscr{C}_0 \oplus \mathscr{C}_1,$$

where $\mathscr{C}_0 = Coker\,L$ and $\mathscr{C}_1 = Im\,L$. The decomposition of an element $w \in C[0,1]$ is denoted by

$$w = w_0 + w_1, w_0 \in \mathscr{C}_0, w_1 \in \mathscr{C}_1.$$

Define the multimap $\Sigma_1 \colon C_{pr}^1[0,1] \to K(C_{pr}^1[0,1])$ by

$$\Sigma_1(x) = P_L(x) + (\Pi_L + K_{P_L, Q_L}) \circ G(x).$$

It is clear that Σ_1 is a completely u.s.c. CJ-multimap and

$$Ind\left(L, G, B_{C_{pr}^1}(0, R')\right) = deg(i - \Sigma_1, B_{C_{pr}^1}(0, R')).$$

Now consider the multimap $\Sigma \colon B_{C_{pr}^1}(0, R') \times [0,1] \to K(C_{pr}^1[0,1])$ given by

$$\Sigma(x, \lambda) = P_L x + (\Pi_L + K_{P_L, Q_L}) \circ \varphi(G(x), \lambda),$$

where the map $\varphi \colon C[0,1] \times [0,1] \to C[0,1]$ is defined as

$$\varphi(w, \lambda) = w_0 + \lambda w_1, \ w = w_0 + w_1, w_0 \in \mathscr{C}_0, w_1 \in \mathscr{C}_1. \tag{5.21}$$

Assume that there exists $(x, \lambda) \in \partial B_{C_{pr}^1}(0, R') \times [0,1]$ such that $x \in \Sigma(x, \lambda)$. Then there is $w \in G(x)$ such that

$$x = P_L x + (\Pi_L + K_{P_L, Q_L}) \circ \varphi(w, \lambda)$$

or equivalently

$$\begin{cases} x' = \lambda w_1 \\ 0 = w_0. \end{cases}$$

If $\lambda \neq 0$ then

$$\int_0^1 V'(x(t))w(t)dt = \frac{1}{\lambda} \int_0^1 V'(x(t))x'(t)dt = \frac{1}{\lambda}\big(V(x(1)) - V(x(0))\big) = 0,$$

giving a contradiction.

If $\lambda = 0$ then $x' = 0$, i.e., $x \equiv a \in \mathbf{R}$. Since $\|a\|_2 = |a| = R' > N$, we have that

$$0 < \int_0^1 V'(a)\tilde{w}(s)ds = V'(a)\Pi_L(\tilde{w}) \tag{5.22}$$

for all $\tilde{w} \in G(a)$. In particular, $0 < V'(a)\Pi_L(w) = V'(a)\Pi_L(w_0) = 0$, that is a contradiction again.

Thus Σ is a homotopy joining the multimaps $\Sigma(\cdot, 1) = \Sigma_1$ and

$$\Sigma(\cdot, 0) = P_L + (\Pi_L + K_{P_L, Q_L}) \circ \varphi(G, 0) = P_L + \Pi_L G. \tag{5.23}$$

The homotopy invariantness property of the topological degree implies that

$$deg(i - \Sigma_1, B_{C_{pr}^1}(0, R')) = deg(i - P_L - \Pi_L G, B_{C_{pr}^1}(0, R')).$$

Notice that the multimap $P_L + \Pi_L G$ has values in $\mathbf{R}$, so

$$deg(i - P_L - \Pi_L G, B_{C_{pr}^1}(0, R')) = deg(i - P_L - \Pi_L G, [-R', R']).$$

In $\mathbf{R}$ the vector multifield $i - P_L - \Pi_L G$ has the form:

$$i - P_L - \Pi_L G = -\Pi_L G.$$

From (5.22) it follows that the fields V' and $\Pi_L G$ are homotopic on $\partial[-R', R']$, so

$$deg(-\Pi_L G, [-R', R']) = -deg(V', [-R', R']) = -Ind\, V.$$

Therefore, $Ind\,(f, G, B_{C^1}(0, R')) \neq 0$, and hence, problem (5.11)–(5.14) has a solution. $\square$

5.3.2   Example

We consider the following problem

$$(x^{2n}(t) + 1)x'(t) - a(t)x^{2n+1}(t) - \frac{x^n(t)x'^{2n}(t)}{n(1 + x'^{2n}(t))} = \mu x(t) + \tilde{B}(t, x(t), y(t)), \tag{5.24}$$

$$y'(t) \in C(t, x(t), y(t)), \tag{5.25}$$

$$x(0) = x(1), \tag{5.26}$$

$$y(0) = y_0, \tag{5.27}$$

where $n \in \mathbf{N}$ is an odd number; $\mu > 0$, $y_0 \in \mathbf{R}$; $a \in C[0, 1]$ is a positive function; $\tilde{B}: [0, 1] \times \mathbf{R} \times \mathbf{R} \to \mathbf{R}$ is a continuous map satisfying (B) and an upper Carathéodory multimap $C: [0, 1] \times \mathbf{R} \times \mathbf{R} \to Kv(\mathbf{R})$ satisfies $(C1)$–$(C2)$.

Theorem 5.11. *For each $\mu > c(1 + de^d)$ problem (5.24)–(5.27) has a solution.*

Proof. Let $A: [0, 1] \times \mathbf{R} \times \mathbf{R} \to \mathbf{R}$ and $B: [0, 1] \times \mathbf{R} \times \mathbf{R} \to \mathbf{R}$ be continuous maps defined by

$$A(t, u, v) = (u^{2n} + 1)v - a(t)u^{2n+1} - \frac{u^n v^{2n}}{n(1 + v^{2n})},$$

$$B(t, u, v) = \mu u + \tilde{B}(t, u, v).$$

Then problem (5.24)–(5.27) can be rewritten in the form of problem (5.11)–(5.14). It is clear that the map B is continuous and satisfies condition (B). We have

$$A_u'(t, u, v) = 2nvu^{2n-1} - (2n + 1)a(t)u^{2n} - \frac{u^{n-1} v^{2n}}{(1 + v^{2n})}, \quad and$$

$$A_v'(t, u, v) = 1 + u^{2n} - \frac{2u^n v^{2n-1}}{(1 + v^{2n})^2}.$$

Notice that $A_u'(t, u, v)$ and $A_v'(t, u, v)$ are continuous and $A_v'(t, 0, 0) = 1$ for all $t \in [0, 1]$. The following estimation holds:

$$
\begin{aligned}
A_v'(t, u, v) &= \frac{(1 + u^{2n})(1 + v^{2n})^2 - 2u^n v^{2n-1}}{(1 + v^{2n})^2} \\
&= \frac{1 + u^{2n} + 2v^{2n} + 2u^{2n}v^{2n} + v^{4n} + u^{2n}v^{4n} - 2u^n v^{2n-1}}{(1 + v^{2n})^2} \\
&= \frac{(u^n - v^{2n-1})^2 + v^{4n} - v^{4n-2} + 1 + 2v^{2n}(1 + u^{2n}) + u^{2n}v^{4n}}{(1 + v^{2n})^2} > 0,
\end{aligned}
$$

since $v^{4n} - v^{4n-2} + 1 > 0$ for all $v \in \mathbf{R}$. So, conditions $(A1)$–$(A2)$ hold.

For every $w \in \mathbf{R}$ and $\lambda \in [0, 1]$, consider equation $(1 - \lambda)v + \lambda A(t, u, v) = w$, or equivalently,

$$v = \frac{w}{1 + \lambda u^{2n}} + \frac{\lambda a(t)u^{2n+1}}{1 + \lambda u^{2n}} + \frac{\lambda u^n v^{2n}}{n(1 + \lambda u^{2n})(1 + v^{2n})}.$$

We see that $|v| < |w| + a(t)|u| + 1$. Set $\alpha(t) = \max\{a(t), 1\}$. Then we have

$$|v| < \alpha(t)(1 + |w| + |u|)$$

and so condition $(A3)$ holds.

Now we prove that the map $V: \mathbf{R} \to \mathbf{R}$, $V(x) = \frac{1}{2}x^2$, is a guiding function for problem (5.24)–(5.27). Since n is an odd number, for every $x \in C_{pr}^1[0, 1]$ the following relation holds:

$$\int_0^1 A(t, x(t), x'(t)) V'(x(t)) dt = \int_0^1 (1 + x^{2n}(t)) x'(t) x(t) dt$$

$$- \int_0^1 a(t) x^{2n+2}(t) dt - \int_0^1 \frac{x^{n+1}(t) x'^{2n}(t)}{n(1 + x'^{2n}(t))} dt \le 0.$$

Choosing an arbitrary $y \in \Pi(x)$ and applying (B), (5.18) we have

$$\int_0^1 B(t, x(t), y(t)) V'(x(t)) dt = \mu \|x\|_2^2 + \int_0^1 \tilde{B}(t, x(t), y(t)) x(t) dt$$

$$\ge \mu \|x\|_2^2 - \int_0^1 c|x(t)|(1 + |x(t)| + |y(t)|) dt$$

$$\ge (\mu - c) \|x\|_2^2 - c\|x\|_2 - c \int_0^1 |x(t)|(|y_0| + d + d\|x\|_2) e^d dt$$

$$\ge (\mu - c - cde^d) \|x\|_2^2 - c(1 + |y_0|e^d + de^d) \|x\|_2 > 0,$$

if

$$\|x\|_2 > \frac{c(1 + |y_0|e^d + de^d)}{\mu - c(1 + de^d)}.$$

From Theorem 5.10 and the fact that $Ind\ V = 1$ it follows that problem (5.24)–(5.27) has a solution.   □

5.4 Global Bifurcation Problem

By applying the oriented coincidence index for a pair consisting of a nonlinear Fredholm operator and a CJ-multimap, we prove a global bifurcation theorem for solutions of families of inclusions with such maps. Using the MGF and the abstract bifurcation result we study the qualitative behavior of branches of periodic trajectories for a feedback control system.

5.4.1 Abstract Result

Let E, E' be Banach spaces. Consider the following family of inclusions

$$f(x, \mu) \in G(x, \mu), \tag{5.28}$$

where $f: E \times \mathbf{R} \to E'$ and $G: E \times \mathbf{R} \to K(E')$.

We assume that the following conditions hold:

(H1) $f \in \Phi_1 C^1(E \times \mathbf{R}, E')$ and the Fredholm structure $\{E \times \mathbf{R}, f\}_\Phi$ on $E \times \mathbf{R}$ induced by f is orientable.

(H2) The restriction $f_{|\Omega}$ is proper for every closed bounded subset $\Omega \subset E \times \mathbf{R}$.

(H3) $G: E \times \mathbf{R} \rightarrow K(E')$ is a completely u.s.c. CJ-multimap and $f(0, \mu) \in G(0, \mu)$ for all $\mu \in \mathbf{R}$.

(H4) There are $\mu_0 \in \mathbf{R}$ and $\varepsilon_0 > 0$ such that for every $\mu \in \mathbf{R}, 0 < |\mu - \mu_0| \le \varepsilon_0$, there exists $\delta_\mu > 0$ which continuously depends on μ and

$$Coin\big(f(\cdot, \mu), G(\cdot, \mu)\big) \cap B_E(0, \delta_\mu) = \{(0, \mu)\}.$$

Denote by $\mathscr{S}$ the set of all nontrivial solutions of (5.28), i.e.,

$$\mathscr{S} = \{(x, \mu) \in E \times \mathbf{R}: x \ne 0 \, and \, f(x, \mu) \in G(x, \mu)\}.$$

We intend to study the global structure of the set $\mathscr{S}$.

Set $\tilde{f}: E \times \mathbf{R} \rightarrow E' \times \mathbf{R}$, $\tilde{f}(x, \mu) = (f(x, \mu), 0)$, and for each $r, \varepsilon > 0$ define the multimap

$$G_r: \overline{U}_{r,\varepsilon} \rightarrow K(E' \times \mathbf{R}),$$

$$G_r(x, \mu) = \{G(x, \mu), r^2 - \|x\|^2\},$$

where

$$\overline{U}_{r,\varepsilon} = \{(x, \mu) \in E \times \mathbf{R}: \|x\|^2 + (\mu - \mu_0)^2 \le r^2 + \varepsilon^2\}.$$

It is easy to see that $\tilde{f}$ is a Fredholm operator of zero index whose restriction to each closed bounded set $\Omega \subset E \times \mathbf{R}$ is proper, and G_r is a compact CJ-multimap. Moreover, the induced by $\tilde{f}$ Fredholm structure on $E \times \mathbf{R}$ is orientable.

Let $0 < \varepsilon < \varepsilon_0$ and $0 < r < \min\{\delta_{\mu_0-\varepsilon}, \delta_{\mu_0+\varepsilon}\}$. We claim that $\tilde{f}$, G_r and $\overline{U}_{r,\varepsilon}$ form a compact triplet. In fact, it is sufficient to verify only that

$$Coin(\tilde{f}, G_r) \cap \partial U_{r,\varepsilon} = \emptyset.$$

To the contrary, assume that there exists $(x, \mu) \in \partial U_{r,\varepsilon}$ such that $\tilde{f}(x, \mu) \in G_r(x, \mu)$. Then $\|x\| = r$ and $f(x, \mu) \in G(x, \mu)$. Since $(x, \mu) \in \partial U_{r,\varepsilon}$ we have that $\mu = \mu_0 \pm \varepsilon$. From the choice of r and (H4) it follows that

$$Coin\big(f(\cdot, \mu_0 \pm \varepsilon), G(\cdot, \mu_0 \pm \varepsilon)\big) \cap B_E(0, r)) = \{(0, \mu_0 \pm \varepsilon)\},$$

giving a contradiction.

Generalizing the Ize's lemma (see, e.g., [77,78,113]) we introduce the following notion.

Definition 5.13. A global bifurcation index of family (5.28) at the point $(0, \mu_0)$ is defined as

$$Bi(0, \mu_0) = Ind(\tilde{f}, G_r, \overline{U}_{r,\varepsilon}). \qquad (5.29)$$

From the properties of the coincidence index it easily follows that a global bifurcation index is well defined, i.e., it does not depend on the choice of ε and r.

Theorem 5.12. *Let conditions $(H1)$–$(H4)$ hold. Assume that $Bi(0, \mu_0) \neq 0$. Then there exists a connected subset $\mathscr{R} \subset \mathscr{S}$ such that $(0, \mu_0) \in \overline{\mathscr{R}}$ and either $\mathscr{R}$ is unbounded or $\overline{\mathscr{R}} \ni (0, \mu_*)$ for some $\mu_* \neq \mu_0$.*

Proof. Let $\mathcal{O} \subset E \times \mathbf{R}$ be an open set defined as

$$\mathcal{O} = (E \times \mathbf{R}) \setminus (\{0\} \times (\mathbf{R} \setminus (\mu_0 - \varepsilon_0, \mu_0 + \varepsilon_0))).$$

From $Ind(\tilde{f}, G_r, \overline{U}_{r,\varepsilon}) \neq 0$ it follows that $(0, \mu_0)$ is a bifurcation point. Let us denote by $\mathscr{W} \subset \mathscr{S} \cup \{(0, \mu_0)\} \subset \mathcal{O}$ the connected component of $(0, \mu_0)$. Assume that $\mathscr{W}$ is compact. Then there exists an open bounded subset $U \subset \mathcal{O}$ such that

$$\overline{U} \subset \mathcal{O}, \ \mathscr{W} \subset U \text{ and } \partial U \cap \mathscr{S} = \emptyset.$$

Hence, $Coin(\tilde{f}, G_r) \cap \partial U = \emptyset$ for each $r > 0$. Further, for any $r, R > 0$, the compact triplets $(\tilde{f}, G_r, \overline{U})$ and $(\tilde{f}, G_R, \overline{U})$ can be joined by the homotopy

$$(\tilde{f}^*, G_{\lambda r + (1-\lambda)R}, \overline{U} \times [0, 1]),$$

where $\tilde{f}^*(x, \mu, \lambda) = \tilde{f}(x, \mu)$ (notice that the Fredholm structure is invariant under this homotopy). For sufficiently large R, $Coin(\tilde{f}, G_R) \cap \overline{U} = \emptyset$, so $Ind(\tilde{f}, G_R, \overline{U}) = 0$. Therefore, $Ind(\tilde{f}, G_r, \overline{U}) = 0$ for every $r > 0$. Let $\Lambda = \{\mu \in \mathbf{R}: (0, \mu) \in \overline{U}\}$. From $\overline{U} \subset \mathcal{O}$ it follows that

$$\Lambda \subset (\mu_0 - \varepsilon_0, \mu_0 + \varepsilon_0). \qquad (5.30)$$

From the continuous dependence of δ_μ on μ it follows that we can choose $0 < \varepsilon < \varepsilon_0$ and $0 < r < \min\{\delta_{\mu_0-\varepsilon}, \delta_{\mu_0+\varepsilon}\}$ such that $\overline{U}_{r,\varepsilon} \subset U$ and

$$Coin(f(\cdot, \mu), G(\cdot, \mu)) \cap B_E(0, r) = \{(0, \mu)\},$$

for all $\mu \in [\mu_0 - \varepsilon_0, \mu_0 + \varepsilon_0] \setminus (\mu_0 - \varepsilon, \mu_0 + \varepsilon)$. From (5.30) and the choice of r, ε (we can take r, ε sufficiently small) we have $Coin(\tilde{f}, G_r) \cap \overline{U} \subset U_{r,\varepsilon}$. So, we obtain

$$0 = Ind(\tilde{f}, G_r, \overline{U}) = Ind(\tilde{f}, G_r, \overline{U}_{r,\varepsilon}) \neq 0,$$

that is a contradiction. Thus, $\mathscr{W}$ is a non-compact component, i.e., either $\mathscr{W}$ is unbounded or $\mathscr{W} \cap \overline{\mathcal{O}} \neq \emptyset$. $\qquad \square$

5.4.2 Global Bifurcation for Families of Periodic Trajectories

We consider, here, the global bifurcation of trajectories for the following one-parameter family of control systems:

$$A\big(t, x(t), x'(t), \mu\big) = B\big(t, x(t), y(t), \mu\big), \ \textit{for all } t \in [0, 1], \tag{5.31}$$

$$y'(t) \in C\big(t, x(t), y(t), \mu\big), \ \textit{for a.a. } t \in [0, 1], \tag{5.32}$$

$$x(0) = x(1), \tag{5.33}$$

$$y(0) = 0, \tag{5.34}$$

where $A: [0, 1] \times \mathbf{R} \times \mathbf{R} \times \mathbf{R} \to \mathbf{R}$, $B: [0, 1] \times \mathbf{R} \times \mathbf{R} \times \mathbf{R} \to \mathbf{R}$ are continuous maps; $C: [0, 1] \times \mathbf{R} \times \mathbf{R} \times \mathbf{R} \to Kv(\mathbf{R})$ is an upper Carathéodory multimap; $\mu \in \mathbf{R}$.

Assume that:

$(A1)'$ There exist continuous partial derivatives $A_u'(t, u, v, \mu)$, $A_v'(t, u, v, \mu)$, $A_\mu'(t, u, v, \mu)$ and moreover, $A_v'(t, u, v, \mu) \neq 0$.

$(A2)'$ $A_v'(t, 0, 0, \mu) = 1$ for all $t \in [0, 1]$ and $\mu \in \mathbf{R}$.

(AB) $A(t, 0, 0, \mu) = B(t, 0, v, \mu)$ for all $(t, v, \mu) \in [0, 1] \times \mathbf{R} \times \mathbf{R}$.

$(C1)'$ The multimap C is uniformly continuous in the second and fourth arguments in the following sense: for each $\varepsilon > 0$ there exists $\delta > 0$ such that

$$C(t, \overline{u}, v, \overline{\mu}) \subset O_\varepsilon\big(C(t, u, v, \mu)\big) \ \textit{for all } (t, v) \in [0, 1] \times \mathbf{R},$$

whenever $|\overline{u} - u| < \delta$ and $|\overline{\mu} - \mu| < \delta$.

$(C2)'$ There exists a constant $d > 0$ such that

$$\|C(t, u, v, \mu)\| \leq d(|\mu| + |u| + |v|),$$

for all $(u, v, \mu) \in \mathbf{R} \times \mathbf{R} \times \mathbf{R}$ and a.a. $t \in [0, 1]$.

As above, for every $(x, \mu) \in C[0, 1] \times \mathbf{R}$ the set $\Pi(x, \mu)$ of all solutions of inclusion (5.32) with initial condition (5.34) is an R_δ set in $C[0, 1]$. Define the multimap

$$\hat{\Pi}: C[0, 1] \times \mathbf{R} \to K\big(C[0, 1] \times C[0, 1] \times \mathbf{R}\big),$$

$$\hat{\Pi}(x, \mu) = \{x\} \times \Pi(x, \mu) \times \{\mu\},$$

and the map

$$\hat{B}: C[0, 1] \times C[0, 1] \times \mathbf{R} \to C[0, 1],$$

$$\hat{B}(x, y, \mu)(t) = B\big(t, x(t), y(t), \mu\big).$$

Then we can rewrite problem (5.31)–(5.34) in the form of the following family of inclusions

$$f(x, \mu) \in G(x, \mu),\tag{5.35}$$

where $G : C_{pr}^1[0, 1] \times \mathbf{R} \to K(C[0, 1])$, $G(x, \mu) = \hat{B} \circ \hat{\Pi}(x, \mu)$, and

$$f : C_{pr}^1[0, 1] \times \mathbf{R} \to C[0, 1],$$

$$f(x, \mu)(t) = A(t, x(t), x'(t), \mu),\tag{5.36}$$

It is easy to see that G is a completely u.s.c. CJ-multimap and f is a Fredholm operator if index 1. From (AB) it follows that $(0, \mu)$ is a solution of (5.35) for every $\mu \in \mathbf{R}$. We call these solutions, trivial solutions. Let us denote by $\mathscr{S}$ the set of all non-trivial solutions of inclusion (5.35), i.e.,

$$\mathscr{S} = \{(x, \mu) \in C_{pr}^1[0, 1] \times \mathbf{R} : x \neq 0 \text{ and } f(x, \mu) \in G(x, \mu)\}.$$

Definition 5.14. A continuously differentiable function $V(u, \mu) : \mathbf{R} \times \mathbf{R} \to \mathbf{R}$ is said to be a local integral guiding function at the point $(0, \mu_0)$, $\mu_0 \in \mathbf{R}$, for problem (5.31)–(5.34), if there exist $\varepsilon_0 > 0$ and a continuous function $\delta : \mathbf{R} \to [0, \infty)$ such that for each μ, $0 < |\mu - \mu_0| \le \varepsilon_0$, the value $\delta_\mu = \delta(\mu) > 0$ and for every $x \in C_{pr}^1[0, 1]$ such that $0 < \|x\|_2 \le \delta_\mu$ the following relations hold:

$$\int_0^1 A\big(t, x(t), x'(t), \mu\big) V_u'\big(x(t), \mu\big) \, dt \le 0,$$

$$\int_0^1 B\big(t, x(t), y(t), \mu\big) V_u'\big(x(t), \mu\big) \, dt > 0$$

for all $y \in \Pi(x, \mu)$.

From the above definition it follows that for every $0 < \varepsilon < \varepsilon_0$ and

$$0 < r < \min\{\delta_{\mu_0-\varepsilon}, \delta_{\mu_0+\varepsilon}\},$$

the vector field

$$V^\sharp : \mathbf{R} \times \mathbf{R} \to \mathbf{R} \times \mathbf{R},$$

$$V^\sharp(u, \mu) = \{-V_u'(u, \mu), \varepsilon^2 - (\mu - \mu_0)^2\},$$

has no zeros on $\partial \overline{U}_{r,\varepsilon}^0$, where

$$\overline{U}_{r,\varepsilon}^0 = \{(u, \mu) \in \mathbf{R} \times \mathbf{R} : u^2 + (\mu - \mu_0)^2 \le r^2 + \varepsilon^2\}.\tag{5.37}$$

Therefore, the topological degree $deg(V^\sharp, \overline{U}_{r,\varepsilon}^0)$ is well defined and does not depend on $(\varepsilon, r) \in (0, \varepsilon_0) \times \big(0, \min\{\delta_{\mu_0-\varepsilon}, \delta_{\mu_0+\varepsilon}\}\big)$. This number we denote by *ind* $V^\sharp$.

Theorem 5.13. *Let conditions* $(A1)'-(A2)'$, (AB) *and* $(C1)'-(C2)'$ *hold. Assume that there exists a local integral guiding function* V *at the point* $(0, \mu_0)$, $\mu_0 \in \mathbf{R}$, *for problem* (5.31)–(5.34) *such that* $ind\, V^\sharp \neq 0$. *Then there exists a connected subset* $\mathscr{R} \subset \mathscr{S}$ *such that* $(0, \mu_0) \in \overline{\mathscr{R}}$ *and at least one of the following occurs:*

(i) $\mathscr{R}$ *is unbounded;*
(ii) $(0, \mu_*) \in \overline{\mathscr{R}}$ *for some* $\mu_* \neq \mu_0$.

Proof. STEP 1. Set $\tilde{f} : C^1_{pr}[0, 1] \times \mathbf{R} \to C[0, 1] \times \mathbf{R}$, $\tilde{f}(x, \mu) = (f(x, \mu), 0)$, and for each $r, \varepsilon > 0$ define the multimap

$$G_r : \overline{U}_{r,\varepsilon} \to K\big(C[0, 1] \times \mathbf{R}\big),$$

$$G_r(x, \mu) = \{G(x, \mu), r^2 - \|x\|^2_{C^1}\},$$

where

$$\overline{U}_{r,\varepsilon} = \{(x, \mu) \in C^1_{pr}[0, 1] \times \mathbf{R} : \|x\|^2_{C^1} + (\mu - \mu_0)^2 \leq r^2 + \varepsilon^2\}.$$

It is easy to see that $\tilde{f}$ is a Fredholm operator of zero index whose restriction to each closed bounded set $\Omega \subset C^1_{pr}[0, 1] \times \mathbf{R}$ is proper, and G_r is a compact CJ-multimap.

Choose arbitrarily $\varepsilon \in (0, \varepsilon_0)$ and $0 < r < \min\{\delta_{\mu_0-\varepsilon}, \delta_{\mu_0+\varepsilon}\}$, where ε_0 is the constant from Definition 5.14. We show that $Coin(\tilde{f}, G_r) \cap \partial U_{r,\varepsilon} = \emptyset$.

To the contrary, assume that there exists $(x, \mu) \in \partial U_{r,\varepsilon}$ such that

$$\tilde{f}(x, \mu) \in G_r(x, \mu).$$

Then there is $y \in \Pi(x, \mu)$ such that

$$A(t, x(t), x'(t), \mu) = B(t, x(t), y(t), \mu) \text{ for all } t \in [0, 1], \text{ and} \tag{5.38}$$

$$\|x\|_{C^1} = r. \tag{5.39}$$

From the fact that $(x, \mu) \in \partial U_{r,\varepsilon}$ and (5.39) it follows that $\mu = \mu_0 \pm \varepsilon$. Moreover, $0 < \|x\|_2 \leq \|x\|_{C^1}$. From the choice of r, Definition 5.14 and (5.38) it follows that

$$0 \geq \int_0^1 A\big(t, x(t), x'(t), \mu\big) V'_u\big(x(t), \mu\big) dt = \int_0^1 B\big(t, x(t), y(t), \mu\big) V'_u\big(x(t), \mu\big) dt > 0,$$

giving a contradiction. So, $(\tilde{f}, G_r, \overline{U}_{r,\varepsilon})$ is a compact triplet.

STEP 2. For given numbers r, ε, where $\varepsilon \in (0, \varepsilon_0)$ and $0 < r < \min\{\delta_{\mu_0-\varepsilon}, \delta_{\mu_0+\varepsilon}\}$ we evaluate the index $Ind(\tilde{f}, G_r, \overline{U}_{r,\varepsilon})$. To this aim, we consider the triplet $(\tilde{f}^\sharp, G_r^\sharp, \overline{U}_{r,\varepsilon} \times [0, 1])$, where

$$\tilde{f}^{\sharp}: C^1_{pr}[0, 1] \times \mathbf{R} \times [0, 1] \to C[0, 1] \times \mathbf{R},$$

$$\tilde{f}^{\sharp}(x, \mu, \lambda) = (f^{\sharp}(x, \mu, \lambda), 0)$$

the map $f^{\sharp}$ is defined analogously as in (5.20), and

$$G^{\sharp}_r: \overline{U}_{r,\varepsilon} \times [0, 1] \to C[0, 1] \times \mathbf{R},$$

$$G^{\sharp}_r(x, \mu, \lambda) = G_r(x, \mu).$$

It is clear that $\tilde{f}^{\sharp}$ is a Fredholm operator of index 1, and $G^{\sharp}_r$ is a compact CJ-multimap.

Assume that there is $(x, \mu, \lambda) \in \partial \overline{U}_{r,\varepsilon} \times [0, 1]$ such that

$$\tilde{f}^{\sharp}(x, \mu, \lambda) \in G^{\sharp}_r(x, \mu, \lambda).$$

Then $\|x\|_{C^1} = r$ and there exists $y \in \Pi(x, \mu)$ such that

$$A^{\sharp}(t, x(t), x'(t), \mu, \lambda) = B(t, x(t), y(t), \mu), \text{ for all } t \in [0, 1].$$

From $\|x\|_{C^1} = r$ it follows that $\mu = \mu_0 \pm \varepsilon$ and $\|x\|_2 \leq r < \min\{\delta_{\mu_0-\varepsilon}, \delta_{\mu_0+\varepsilon}\}$. Therefore

$$0 < \int_0^1 B(t, x(t), y(t), \mu) V'_u(x(t), \mu) \, dt = \int_0^1 A^{\sharp}(t, x(t), x'(t), \mu, \lambda) V'_u(x(t), \mu) \, dt$$

$$= \lambda \int_0^1 A(t, x(t), x'(t), \mu) V'_u(x(t), \mu) \, dt \leq 0,$$

giving a contradiction.

Thus, $(\tilde{f}^{\sharp}, G^{\sharp}_r, \overline{U}_{r,\varepsilon} \times [0, 1])$ is a compact homotopy joining the triplets $(\tilde{f}, G_r, \overline{U}_{r,\varepsilon})$ and $(\tilde{L}, G_r, \overline{U}_{r,\varepsilon})$, where $\tilde{L}(x, \mu) = (x', 0)$. By Theorem 5.5 we obtain

$$| \operatorname{Ind}(\tilde{f}, G_r, \overline{U}_{r,\varepsilon}) | = | \operatorname{Ind}(\tilde{L}, G_r, \overline{U}_{r,\varepsilon}) |.$$

Set $\tilde{\Sigma}: \overline{U}_{r,\varepsilon} \times [0, 1] \to C^1_{pr}[0, 1] \times \mathbf{R}$,

$$\tilde{\Sigma}(x, \mu, \lambda) = \{x - P_L x - (\Pi_L + K_{P_L, Q_L}) \circ \varphi(G(x, \mu), \lambda), \tau\},$$

$$\tau = \lambda(\|x\|^2_{C^1} - r^2) + (1 - \lambda)(\varepsilon^2 - (\mu - \mu_0)^2),$$

where the map φ was defined in (5.21) and the maps P_L, Π_L, K_{P_L, Q_L} are mentioned in the previous section.

It is easy to see that $\tilde{\Sigma}$ is a compact CJ-multifield. Assume that there exists $(x, \mu, \lambda) \in \partial U_{r,\varepsilon} \times [0, 1]$ such that $0 \in \tilde{\Sigma}(x, \mu, \lambda)$. Then

$$\lambda(\|x\|_{C^1}^2 - r^2) + (1 - \lambda)(\varepsilon^2 - (\mu - \mu_0)^2) = 0 \qquad (5.40)$$

and there is $w \in G(x, \mu)$ such that

$$\begin{cases} x' = \lambda w_1 \\ 0 = w_0, \end{cases} \qquad (5.41)$$

where $w = w_0 + w_1$, $w_0 \in \mathscr{C}_0$, $w_1 \in \mathscr{C}_1$.

From (5.40) and the fact that $(x, \mu) \in \partial U_{r,\varepsilon}$ it follows that

$$\|x\|_{C^1} = r \ and \ \mu = \mu_0 \pm \varepsilon. \qquad (5.42)$$

If $\lambda > 0$ then from the choice of r we have

$$\int_0^1 V_u'(x(t), \mu) w(t) dt = \frac{1}{\lambda} \int_0^1 V_u'(x(t), \mu) x'(t) dt$$

$$= \frac{1}{\lambda} \big(V(x(1), \mu) - V(x(0), \mu) \big) = 0,$$

giving a contradiction.

If $\lambda = 0$ then $x' = 0$, i.e., $x \equiv a \in \mathbf{R}$. Since $|a| = r < \min\{\delta_{\mu_0-\varepsilon}, \delta_{\mu_0+\varepsilon}\}$, we have that

$$0 < \int_0^1 V_u'(a, \mu) \tilde{w}(s) ds = V_u'(a, \mu) \Pi_L(\tilde{w}) \qquad (5.43)$$

for all $\tilde{w} \in G(a, \mu)$. In particular, $0 < V_u'(a, \mu) \Pi_L(w) = V_u'(a, \mu) \Pi_L(w_0) = 0$. That is a contradiction again.

So, $\tilde{\Sigma}$ is a homotopy, and hence

$$Ind\,(\tilde{L}, G_r, \overline{U}_{r,\varepsilon}) = deg\big(\tilde{\Sigma}(\cdot, \cdot, 0), \overline{U}_{r,\varepsilon} \big).$$

Similarly to (5.23), the multimap $\tilde{\Sigma}(\cdot, \cdot, 0)$ has the form:

$$\tilde{\Sigma}(\cdot, \cdot, 0) = \{i - P_L - \Pi_L G, \varepsilon^2 - (\mu - \mu_0)^2\},$$

Using the homotopy invariance property of the topological degree we obtain that

$$Ind\,(\tilde{L}, G_r, \overline{U}_{r,\varepsilon}) = deg(\{i - P_L - \Pi_L G, \varepsilon^2 - (\mu - \mu_0)^2\}, \overline{U}_{r,\varepsilon}).$$

Notice that the multimap $P_L + \Pi_L G$ has values in $\mathbf{R}$, so

$$deg\big(\{i - P_L - \Pi_L G, \varepsilon^2 - (\mu - \mu_0)^2\}, \overline{U}_{r,\varepsilon}\big)$$

$$= deg\big(\{i - P_L - \Pi_L G, \varepsilon^2 - (\mu - \mu_0)^2\}, \overline{U}^0_{r,\varepsilon}\big)$$

$$= deg\big(\{-\Pi_L G, \varepsilon^2 - (\mu - \mu_0)^2\}, \overline{U}^0_{r,\varepsilon}\big),$$

where $\overline{U}^0_{r,\varepsilon}$ is defined in (5.37).

From (5.43) it follows that the vector fields $\{-\Pi_L G, \varepsilon^2 - (\mu - \mu_0)^2\}$ and $V^\sharp$ are homotopic on $\partial U^0_{r,\varepsilon}$. Therefore, $Ind\,(\tilde{f}, G_r, \overline{U}_{r,\varepsilon}) \neq 0$.

So, in accordance with (5.29), the global bifurcation index $Bi\,(0, \mu_0)$ of family of inclusions (5.35) is non-zero and we can apply Theorem 5.12. $\qquad\square$

5.4.3 Example

Consider system (5.31)–(5.34) with the given maps

$$A(t, u, v, \mu) = (\mu^{2n} u^{2n} + 1)v - \mu a(t) u^{2n+1} - \frac{\mu^n u^n v^{2n}}{n(1 + v^{2n})}, \quad \text{and}$$

$$B(t, u, v, \mu) = u\big(T\mu + b(t)v\big),$$

where $n \in \mathbf{N}$ is an odd number; $a \in C[0, 1]$ is a positive function; $b \in C[0, 1]$ and $T > 0$.

Theorem 5.14. *Let conditions $(C1)'$–$(C2)'$ hold. In addition, assume that*

$$\|b\|_C < \frac{T}{de^d},$$

where d is the constant from $(C2)'$.
Then there exists a connected subset $\mathscr{R} \subset \mathscr{S}$ such that $(0,0) \in \overline{\mathscr{R}}$ and $\mathscr{R}$ is unbounded.

Proof. At first we claim that the map A satisfies conditions $(A1)'$–$(A2)'$. In fact, for every $(t, u, v, \mu) \in [0, 1] \times \mathbf{R} \times \mathbf{R} \times \mathbf{R}$ we have

$$A'_u(t, u, v, \mu) = 2n\mu^{2n} u^{2n-1} - (2n + 1)\mu a(t) u^{2n} - \frac{\mu^n u^{n-1} v^{2n}}{1 + v^{2n}},$$

$$A'_v(t, u, v, \mu) = \mu^{2n} u^{2n} + 1 - \frac{2\mu^n u^n v^{2n-1}}{(1 + v^{2n})^2},$$

$$A'_\mu(t, u, v, \mu) = 2n\mu^{2n-1} u^{2n} v - a(t) u^{2n+1} - \frac{\mu^{n-1} u^n v^{2n}}{1 + v^{2n}}.$$

It is clear that the maps A'_u, A'_v and A'_μ are continuous and $A'_v(t, 0, 0, \mu) = 1$ for all t, μ.

Moreover,

$$
\begin{aligned}
A'_v(t, u, v, \mu) &= \frac{(1 + \mu^{2n}u^{2n})(1 + v^{2n})^2 - 2\mu^n u^n v^{2n-1}}{(1 + v^{2n})^2} \\
&= \frac{1 + \mu^{2n}u^{2n} + 2v^{2n} + 2\mu^{2n}u^{2n}v^{2n} + v^{4n} + \mu^{2n}u^{2n}v^{4n} - 2\mu^n u^n v^{2n-1}}{(1 + v^{2n})^2} \\
&= \frac{(\mu^n u^n - v^{2n-1})^2 + v^{4n} - v^{4n-2} + 1 + 2v^{2n}(1 + \mu^{2n}u^{2n}) + \mu^{2n}u^{2n}v^{4n}}{(1 + v^{2n})^2} > 0.
\end{aligned}
$$

So, conditions $(A1)'$–$(A2)'$ hold. Condition (AB) is followed immediately from the definition of A and B.

Let $V : \mathbf{R} \times \mathbf{R} \to \mathbf{R}$, $V(u, \mu) = \frac{1}{2}\mu u^2$. Let us show that V is a local integral guiding function at $(0, 0)$ for problem (5.31)–(5.34).

To this aim, we fix $\mu \neq 0$. For every $x \in C^1_{pr}[0, 1]$ we have

$$
\int_0^1 A(t, x(t), x'(t), \mu) V'_\mu(x(t), \mu) dt = \int_0^1 \mu(1 + \mu^{2n} x^{2n}(t)) x'(t) x(t) dt
$$

$$
-\mu^2 \int_0^1 a(t) x^{2n+2}(t) dt - \frac{\mu^{n+1}}{n} \int_0^1 \frac{x^{n+1}(t) x'^{2n}(t)}{1 + x'^{2n}(t)} dt \leq 0. \tag{5.44}
$$

Choose arbitrarily $y \in \Pi(x, \mu)$. Analogously to (5.18) we have

$$
|y(t)| \leq de^d (|\mu| + \|x\|_2),
$$

for all $t \in [0, 1]$.
Therefore,

$$
\begin{aligned}
\int_0^1 B(t, x(t), y(t), \mu) V'_\mu(x(t), \mu) dt &= \mu^2 T \|x\|_2^2 + \mu \int_0^1 b(t) x^2(t) y(t) dt \\
&\geq T\mu^2 \|x\|_2^2 - |\mu| \int_0^1 x^2(t) |b(t)| \, |y(t)| dt \\
&\geq T\mu^2 \|x\|_2^2 - |\mu| \|b\|_C de^d (|\mu| + \|x\|_2) \int_0^1 x^2(t) dt \\
&= \mu^2 (T - \|b\|_C de^d) \|x\|_2^2 - |\mu| \|b\|_C de^d \|x\|_2^3 > 0,
\end{aligned}
\tag{5.45}
$$

provided

$$0 < \|x\|_2 < \frac{T - \|b\|_C de^d}{de^d \|b\|_C} |\mu|. \tag{5.46}$$

For sufficiently small $0 < \varepsilon < \varepsilon_0$ and

$$0 < r < \min\{\delta_{\mu_0 - \varepsilon}, \delta_{\mu_0 + \varepsilon}\},$$

consider the vector multifield

$$V^\sharp : \overline{U}_{r,\varepsilon}^0 \to \mathbf{R} \times \mathbf{R},$$

$$V^\sharp(u, \mu) = \{-\mu u, \varepsilon^2 - \mu^2\},$$

where

$$\overline{U}_{r,\varepsilon}^0 = \{(u, \mu) \in \mathbf{R} \times \mathbf{R} : u^2 + \mu^2 \le r^2 + \varepsilon^2\}.$$

It is easy to see that $V^\sharp(u, \mu) = 0$ at $(0, \varepsilon)$ and $(0, -\varepsilon)$. Moreover, the Fréchet's derivative of $V^\sharp$ at $(0, \mu_*)$ is

$$\left(DV^\sharp(0, \mu_*)\right)(u, \mu) = \{-\mu_* u, -2\mu_* \mu\}.$$

Consequently, $deg(V^\sharp, \overline{U}_{r,\varepsilon}^0) = 2$. Notice that for all $\mu \neq 0$ relations (5.44) and (5.45) hold for $x \in C_{pr}^1[0, 1]$ satisfying (5.46). Hence, $(0, 0)$ is the unique bifurcation point of problem (5.31)–(5.34). Applying Theorem 5.13 we can describe the global structure of the solution set of problem (5.31)–(5.34).   □

References

1. D. Affane, D. Azzam-Laouir, A control problem governed by a second order differential inclusion. Appl. Anal. **88**(12), 1677–1690 (2009)
2. R.P. Agarwal, S.R. Grace, D. O'Regan, Oscillation theorems for second order differential inclusions. Int. J. Dyn. Syst. Differ. Equat. **1**(2), 85–88 (2007)
3. S. Aizicovici, N.H. Pavel, Anti-periodic solutions to a class of nonlinear differential equations in Hilbert space. J. Funct. Anal. **99**(2), 387–408 (1991)
4. J.C. Alexander, P.M. Fitzpatrick, Global bifurcation for solutions of equations involving several parameter multivalued condensing mappings, in *Proceedings of the Fixed Point Theory* (Sherbrooke, QC, 1980), ed. by E. Fadell, G. Fournier. Springer Lecture Notes, vol. 886, pp. 1–19
5. J. Andres, L. Górniewicz, Topological fixed point principles for boundary value problems, in *Topological Fixed Point Theory and Its Applications*, vol. 1 (Kluwer Academic, Dordrecht, 2003)
6. J. Andres, L. Malaguti, M. Pavlačkova, Strictly localized bounding functions for vector second-order boundary value problems. Nonlinear Anal. **71**(12), 6019–6028 (2009)
7. J. Andres, L. Malaguti, M. Pavlačkova, On second-order boundary value problems in Banach spaces: a bound sets approach. Topol. Meth. Nonlinear Anal. **37**(2), 303–341 (2011)
8. J. Andres, L. Malaguti, V. Taddei, Bounded solutions of Carathèodory differential inclusions: a bounded set approach. Abstr. Appl. Math. **9**, 547–571 (2003)
9. J. Andres, L. Malaguti, V. Taddei, A bounding functions approach to multivalued boundary value problems. Dyn. Syst. Appl. **16**, 37–48 (2007)
10. J. Andres, M. Kożuşnìkovà, L. Malaguti, Bound sets approach to boundary value problems for vector second-order differential inclusions. Nolinear Anal. **71**, 28–44 (2009)
11. J. Appell, E. De Pascale, H.T. Nguyen, P.P. Zabreiko, Multi-valued superpositions. Dissertationes Math. CCCXLV, pp. 1–97, 1995
12. E.P. Avgerinos, N.S. Papageorgiou, N. Yannakakis, Periodic solutions for second order differential inclusions with nonconvex and unbounded multifunction. Acta Math. Hung. **83**(4), 303–314 (1999)
13. J.P. Aubin, H. Frankowska, *Set-Valued Analysis* (Birkhauser, Boston, 1990)
14. V. Barbu, *Nonlinear Semigroups and Differential Equations in Banach Spaces* (Noordhoff International Publishing, Leyden, 1976)
15. I. Benedetti, L. Malaguti, V. Taddei, Two-point b.v.p. for multivalued equations with weakly regular r.h.s. Nonlinear Anal. **74**(11), 3657–3670 (2011)
16. P. Benevieri, M. Furi, A simple notion of orientability for Fredholm maps of index zero between Banach manifolds and degree theory. Ann. Sci. Math. Qué. **22**, 131–148 (1998)

17. P. Benevieri, M. Furi, On the concept of orientability for Fredholm maps between real Banach manifolds. Topol. Meth. Nonlinear Anal. **16**, 279–306 (2000)

18. M. Benchohra, S.K. Ntouyas, Controllability of second-order differential inclusions in Banach spaces with nonlocal conditions. J. Optim. Theor. Appl. **107**(3), 559–571 (2000)

19. N.A. Bobylev, The construction of regular guiding functions. Dokl. Akad. Nauk SSSR (Russian) **183**, 265–266 (1968)

20. N.A. Bobylev, Yu.M. Burman, S.K. Korovin, in *Appoximation Procedures in Nonlinear Oscillation Theory*. de Gruyter Series in Nonlinear Analysis and Applications, vol. 2 (Walter de Gruyter, Berlin, 1994)

21. N.A. Bobylev, V.S. Klimov, *Methods of Nonlinear Analysis in Nonsmooth Optimization Problems* (Russian) (Nauka, Moscow, 1992)

22. H. Bohnenblust, S. Karlin, On a theorem of Ville, in *Contributions in the Theory of Games*, vol. 1, ed. by H.W. Kuhn, A.W.Tucker (Princeton University Press, Princeton, 1950), pp. 155–160

23. Yu.G. Borisovich, Topological characteristics and investigation of the solvability of nonlinear problems. Izv. Vyssh. Uchebn. Zaved. Mat. (Russian) (2), 3–23 (1997); English translation: Russ. Math. (Iz. VUZ) **41**(2), 1–21 (1997)

24. Yu.G. Borisovich, B.D. Gelman, A.D. Myshkis, V.V. Obukhovskii, Topological methods in the theory of fixed points of multivalued mappings. Uspekhi Mat. Nauk (Russian) **35**(1)(211), 59–126 (1980); English translation: Russ. Math. Surv. **35**, 65–143 (1980)

25. Yu.G. Borisovich, B.D. Gelman, A.D. Myshkis, V.V. Obukhovskii, *Introduction to the Theory of Multivalued Maps and Differential Inclusions*, 2nd edn. (Librokom, Moscow, 2011) (in Russian)

26. Yu.G. Borisovich, Yu.E. Gliklikh, in *On the Lefschetz Number for a Class of Multi-Valued Maps*. Seventh Math. Summer School (Katsiveli, 1969), pp. 283–294. Izd. Akad. Nauk Ukrain. SSR (Kiev, 1970) (in Russian)

27. Yu.G. Borisovich, V.G. Zvyagin, Yu.I. Sapronov, Nonlinear Fredholm mappings, and Leray-Schauder theory. Uspehi Mat. Nauk (in Russian) **32**(4)(196), 3–54 (1977). English translation: Russ. Math. Surv. **32**(4), 1–54 (1977)

28. Yu.G. Borisovich, V.G. Zvyagin, V.V. Shabunin, On the solvability in W_p^{2m+1} of the nonlinear Dirichlet problem in a narrow strip. Dokl. Akad. Nauk (in Russian) **334**(6), 683–685 (1994). English translation: Russ. Acad. Sci. Dokl. Math. **49**(1), 179–182 (1994)

29. K. Borsuk, in *Theory of Retracts*. Monografie Mat. vol. 44 (PWN, Warszawa, 1967)

30. F.E. Browder, W.V Petryshyn, Approximation methods and the generalized topological degree for nonlinear mappings in Banach spaces. J. Funct. Anal. **3**, 217–245 (1969)

31. C. Castaing, M. Valadier, in *Convex Analysis and Measurable Multifunctions*. Lecture Notes in Mathematics, vol. 580 (Springer, Berlin, 1977)

32. K.C. Chang, The obstacle problem and partial differential equations with discontinuous nonlinearities. Comm. Pure Appl. Math. **33**(2), 117–146 (1980)

33. H.L. Chen, Anti-periodic wavelets. J. Comput. Math. **14**(1), 32–39 (1996)

34. F.H. Clarke, in *Optimization and Nonsmooth Analysis*, 2nd edn. Classics in Applied Mathematics, vol. 5 (Society for Industrial and Applied Mathematics (SIAM), Philadelphia, 1990)

35. Yu. Chena, D. O'Regan, R.P. Agarwal, Anti-periodic solutions for evolution equations associated with monotone type mappings. Appl. Math. Lett. **23**(11), 1320–1325 (2010)

36. J.-F. Couchouron, R. Precup, Anti-periodic solutions for second order differential inclusions. Electron. J. Differ. Equat. **2004**(124), 1–17 (2004)

37. F.S. De Blasi, L. Górniewicz, G. Pianigiani, Topological degree and periodic solutions of differential inclusions. Nonlinear Anal. Theor. Meth. Ser. A **37**(2), 217–243 (1999)

38. K. Deimling, *Nonlinear Functional Analysis* (Springer, Berlin, 1985)

39. K. Deimling, Multivalued Differential Equations, in *De Gruyter Series in Nonlinear Analysis and Applications*, vol. 1 (Walter de Gruyter, Berlin, 1992)

40. V.F. Dem'yanov, L.V. Vasil'ev, *Nondifferentiable Optimization* (Nauka, Moscow, 1981) (in Russian); English translation: Translation Series in Mathematics and Engineering (Optimization Software, Inc., Publications Division, New York, 1985)

41. Z. Denkowski, S. Migòrski, N.S. Papageorgiou, *An Introduction to Nonlinear Analysis: Theory* (Kluwer, Boston, 2003)
42. S. Domachowski, J. Gulgowski, A global bifurcation theorem for convex-valued differential inclusions. Z. Anal. Anwendungen **23**(2), 275–292 (2004)
43. J. Eisner, M. Kuçera, M. Väth, Degree and global bifurcation for elliptic equations with multivalued unilateral conditions. Nonlinear Anal. **64**(8), 1710–1736 (2006)
44. I. Ekland, R. Temam, *Convex Analysis and Variation Problems* (North Holland, Amsterdam, 1979)
45. K.D. Elworthy, A.J. Tromba, Differential structures and Fredholm maps on Banach manifolds, in *Global Analysis*, 1970. Proceedings of the Symposium on Pure Mathematics, vol. XV, Berkeley, CA (American Mathematical Society, Providence, 1968), pp. 45–94
46. L. Erbe, W. Krawcewicz, Boundary value problems for second order nonlinear differential inclusions, in *Qualitative Theory of Differential Equations*, Szeged, 1988. Colloq. Math. Soc. Janos Bolyai, vol. 53 (North-Holland, Amsterdam, 1990), pp. 163–171
47. L. Erbe, W. Krawcewicz, Existence of solutions to boundary value problems for impulsive second order differential inclusions. Rocky Mt. J. Math. **22**(2), 519–539 (1992)
48. M. Feçkan, Bifurcation from homoclinic to periodic solutions in ordinary differential equations with multivalued perturbations. J. Differ. Equat. **130**, 415–450 (1996)
49. M. Feçkan, Bifurcation of periodic solutions in differential inclusions. Appl. Math. **42**(5), 369–393 (1997)
50. M. Feçkan, Bifurcation from homoclinic to periodic solutions in singularly perturbed direrential inclusions. Proc. R. Soc. Edinb. **127A**, 727–753 (1997)
51. M. Feçkan, in *Topological Degree Approach to Bifurcation Problems*. Topological Fixed Point Theory and Its Applications, vol. 5 (Springer, New York, 2008)
52. M. Feçkan, Bifurcation of periodic solutions in forced ordinary differential inclusions. Differ. Equat. Appl. **4**(1), 459–472 (2009)
53. M. Filippakis, L. Gasin'ski, N.S. Papageorgiou, Nonsmooth generalized guiding functions for periodic differential inclusions. NoDEA **13**, 43–66 (2006)
54. P.M. Fitzpatrick, J. Pejsachowicz, P.J. Rabier, Orientability of Fredholm families and topological degree for orientable non-linear Fredholm mappings. J. Funct. Anal. **124**, 1–39 (1994)
55. A. Fonda, Guiding functions and periodic solutions to functional differential equations. Proc. Am. Math. Soc. **99**(1), 79–85 (1987)
56. A. Fryszkowski, in *Fixed Point Theory for Decomposable Sets*. Topological Fixed Point Theory and Its Applications, vol. 2 (Kluwer, Dordrecht, 2004)
57. D. Gabor, The coincidence index for fundamentally contractible multivalued maps with nonconvex values. Ann. Polon. Math. **75**(2), 143–166 (2000)
58. D. Gabor, W. Kryszewski, A coincidence theory involving Fredholm operators of nonnegative index. Topol. Meth. Nonlinear Anal. **15**(1), 43–59 (2000)
59. D. Gabor, W. Kryszewski, A global bifurcation index for set-valued perturbations of Fredholm operators. Nonlinear Anal. **73**(8), 2714–2736 (2010)
60. D. Gabor, W. Kryszewski, Alexander invariant for perturbations of Fredholm operators. Nonlinear Anal. **74**(18), 6911–6932 (2011)
61. G. Gabor, R. Pietkun, Periodic solutions of differential inclusions with retards. Topol. Meth. Nonlinear Anal. **16**, 103–123 (2000)
62. R.E. Gaines, J.L. Mawhin, in *Coincidence Degree and Nonlinear Differential Equations*. Lecture Notes in Mathematics, vol. 568 (Springer, Berlin, 1977)
63. E.A. Gango, A.I. Povolotskii, Proper guiding functions for differential equations with multivalued right-hand side. (in Russian), in *Teoriya funkts. i funktsion. analiz*, (Leningrad. Gos. Ped. Inst., Leningrad, 1975), pp. 35–41
64. L. Górniewicz, in *Topological Fixed Point Theory of Multivalued Mappings*, 2nd edn. Topological Fixed Point Theory and Its Applications, vol. 4 (Springer, Dordrecht, 2006)
65. L. Górniewicz, A. Granas, W. Kryszewski, On the homotopy method in the fixed point index theory of multi-valued mappings of compact absolute neighborhood retracts. J. Math. Anal. Appl. **161**(2), 457–473 (1991)

66. L. Górniewicz, W. Kryszewski, Bifurcation invariants for acyclic mappings. Rep. Math. Phys. **31**(2), 217–239 (1992)

67. L. Górniewicz, S. Plaskacz, Periodic solutions of differential inclusions in $\mathbf{R}^n$. Boll. Un. Mat. Ital. A(7) **7**(3), 409–420 (1993)

68. S.R. Grace, R.P. Agarwal, D. O'Regan, A selection of oscillation criteria for second-order differential inclusions. Appl. Math. Lett. **22**(2), 153–158 (2009)

69. J. Gulgowski, A global bifurcation theorem with applications to nonlinear Picard problems. Nonlinear Anal. **41**, 787–801 (2000)

70. R. Hakl, P.J. Torres, On periodic solutions of secord-order differential equations with attractive-repulsive singularities. J. Differ. Equat. **248**, 111–126 (2010)

71. J.K. Hale, J. Kato, Phase space for retarded equations with infinite delay. Funkcial. Ekvac. **21**(1), 11–41 (1978)

72. Ph. Hartman, in *Ordinary Differential Equations*. Corrected reprint of the second (1982) edition (Birkhauser, Boston). Classics in Applied Mathematics, vol. 38 (Society for Industrial and Applied Mathematics (SIAM), Philadelphia, 2002)

73. Y. Hino, S. Murakami, T. Naito, in *Functional Differential Equations with Infinite Delay*. Lecture Notes in Mathematics, vol. 1473 (Springer, Berlin, 1991)

74. M.W. Hirsch, in *Differential Topology*. Graduate Texts in Mathematics, vol. 33 (Springer, New York, 1994)

75. S. Hu, N.S. Papageorgiou, in *Handbook of Multivalued Analysis. Vol. I. Theory*. Mathematics and Its Applications, vol. 419 (Kluwer, Dordrecht, 1997)

76. D.M. Hyman, On decreasing sequences of compact absolute retracts. Fund. Math. **64**, 91–97 (1969)

77. J. Ize, Bifurcation theory for Fredholm operators. Mem. Am. Math. Soc. **7**(174), viii+128 pp (1976)

78. J. Ize, Topological bifurcation, in *Topological Nonlinear Analysis: Degree, Singularity and Variations*, ed. by M. Matzeu, A. Vignoli. Progress in Nonlinear Differential Equations and Their Applications, vol. 15 (Birkhäuser, Boston, 1995), pp. 341–463

79. J. Jezierski, W. Marzantowicz, in *Homotopy Methods in Topological Fixed and Periodic Point Theory*. Topological Fixed Point Theory and Applications, vol. 2 (Springer, Dordrecht, 2006)

80. M. Kamenskii, V. Obukhovskii, P. Zecca, in *Condensing Multivalued Maps and Semilinear Differential Inclusions in Banach Spaces*. de Gruyter Series in Nonlinear Analysis and Applications, vol. 7 (Walter de Gruyter, Berlin, 2001)

81. I.-S. Kim, Yu.-H. Kim, A global bifurcation for nonlinear inclusions. Nonlinear Anal. **68**(1), 343–348 (2008)

82. S.V. Kornev, V.V. Obukhovskii, On non-smooth multivalent guiding functions. Differ. Uravn. **39**(11), 1497–1502 (2003) (in Russian); English translation: Differ. Equat. **39**(11), 1578–1584 (2003)

83. S.V. Kornev, V.V. Obukhovskii, On some versions of the topological degree theory for nonconvex-valued multimaps. Trudy Mat. Fac. Voronezh Univ. (N.S.) **8**, 56–74 (2004) (in Russian)

84. S. Kornev, V. Obukhovskii, On some developments of the method of integral guiding functions. Funct. Differ. Equat. **12**(3–4), 303–310 (2005)

85. S.V. Kornev, V.V. Obukhovskii, Nonsmooth guiding potentials in problems of forced oscillations, Avtomat. i Telemekh. (1), 3–10 (2007) (in Russian); English translation: Autom. Remote Control **68**(1), 1–8 (2007)

86. S.V. Kornev, V.V. Obukhovskii, Localization of the method of guiding functions in the problem about periodic solutions of differential inclusions. Izvestiya Vysshikh Uchebnykh Zavedenii. Matematika (5), 23–32 (2009) (in Russian). English translation: Russ. Math. **53**(5), 19–27 (2009)

87. A.M. Krasnosel'skii, M.A. Krasnosel'skii, J. Mawhin, Differential inequalities in problems of forced nonlinear oscillations. Nonlinear Anal. **25**(9–10), 1029–1036 (1995)

88. A.M. Krasnosel'skii, M.A. Krasnosel'skii, J. Mawhin, A. Pokrovskii, Generalized guiding functions in a problem on high frequency forced oscillations. Nonlinear Anal. **22**(11), 1357–1371 (1994)

89. M.A. Krasnosel'skii, *Topological Methods in the Theory of Nonlinear Integral Equations* (Gostekhizdat, Moscow, 1956) (in Russian); English translation (A Pergamon Press Book The Macmillan Co., New York, 1964)

90. M.A. Krasnosel'skii, *The Operator of Translation Along the Trajectories of Differential Equations* (Nauka, Moscow, 1966) (in Russian); English translation: Translations of Mathematical Monographs, vol. 19 (American Mathematical Society, Providence, 1968)

91. M.A. Krasnosel'skii, A.I. Perov, On a certain priciple of existence of bounded, periodic and almost periodic solutions of systems of ordinary differential equations. Dokl. Akad. Nauk SSSR **123**(2), 235–238 (1958) (in Russian)

92. M.A. Krasnosel'skii, A.I. Perov, A.I. Povolotskii, P.P. Zabreiko, *Plane Vector Fields* (Gosudarstv. Izdat. Fiz.-Mat. Lit., Moscow, 1963) (in Russian); English translation (Academic, New York, 1966)

93. M.A. Krasnosel'skii, A.V. Pokrovskii, *Systems with Hysteresis* (Nauka, Moscow, 1983) (in Russian); English translation (Springer, Berlin, 1989)

94. M.A. Krasnosel'skii, A.V. Pokrovskii, On elliptic equations with discontinuous nonlinearities (in Russian). Dokl. Akad. Nauk **342**(6), 731–734 (1995)

95. M.A. Krasnosel'skii, P.P. Zabreiko, *Geometrical Methods of Nonlinear Analysis* (Nauka, Moscow, 1975); English translation: Grundlehren der Mathematischen Wissenschaften, vol. 263 (Springer, Berlin, 1984)

96. W. Kryszewski, *Homotopy Properties of Set-Valued Mappings* (Univ. N. Copernicus Publishing, Torun, 1997)

97. S. Kyritsi, N. Matzakos, N.S. Papageorgiou, Periodic problems for strongly nonlinear second-order differential inclusions. J. Differ. Equat. **183**(2), 279–302 (2002)

98. Y.C. Liou, V. Obukhovskii, J.C. Yao, Application of a coincidence index to some classes of impulsive control systems. Nonlinear Anal. **69**(12), 4392–4411 (2008)

99. N.V. Loi, Application of the method of integral guiding functions to bifurcation problems of periodic solutions of differential inclusions. Tambov Univ. Rep. Ser. Nat. Tech. Sci. **14**(4), 738–741 (2009) (in Russian)

100. N.V. Loi, Method of guiding functions for differential inclusions in a Hilbert space. Differ. Uravn. **46**(10), 1433–1443 (2010) (in Russian). English translation: Differ. Equat. **46**(10), 1438–1447 (2010)

101. N.V. Loi, Guiding functions and global bifurcation of periodic solutions of functional differential inclusions with infinite delay. Topol. Meth. Nolinear Anal. **40**, 359–370 (2012)

102. N.V. Loi, V.V. Obukhovskii, On application of the method of guiding functions to bifurcation problem of periodic solutions of differential inclusions. Vestnik Ross. Univ. Dr. Narod. (Russian) **4**, 14–27 (2009)

103. N.V. Loi, V. Obukhovskii, On the global bifurcation for solutions of linear fredholm inclusions with convex-valued perturbations. Fixed Point Theor. **10**(2), 289–303 (2009)

104. N.V. Loi, V. Obukhovskii, On global bifurcation of periodic solutions for functional differential inclusions. Funct. Differ. Equat. **17**(1–2), 157–168 (2010)

105. N.V. Loi, V. Obukhovskii, On the existence of solutions for a class of second-order differential inclusions and applications. J. Math. Anal. Appl. **385**, 517–533 (2012)

106. N.V. Loi, V. Obukhovskii, Guiding functions for generalized periodic problems and applications. Appl. Math. Comput. **218**, 11719–11726 (2012)

107. N.V. Loi, V. Obukhovskii, J.-C. Yao, A bifurcation of solutions of nonlinear Fredholm inclusions involving CJ-multimaps with applications to feedback control systems. Set Valued Var. Anal. (accepted). doi:10.1007/s11228-012-0226-z

108. N.V. Loi, V. Obukhovskii, P. Zecca, Non-smooth guiding functions and periodic solutions of functional differential inclusions with infinite delay in Hilbert spaces. Fixed Point Theor. **13**(2), 565–582 (2012)

109. N.V. Loi, V. Obukhovskii, P. Zecca, On the global bifurcation of periodic solutions of differential inclusions in Hilbert spaces. Nonlinear Anal. **76**, 80–92 (2013)

110. I. Massabo, P. Nistri, A topological degree for mulJvalued A-proper maps in Banach spaces. Boll. U.M.I. **13-B**, 672–685 (1976)

111. J. Mawhin, in *Topological Degree Methods in Nonlinear Boundary Value Problems.*
     Expository Lectures from the CBMS Regional Conference Held at Harvey Mudd College,
     Claremont, CA, June 9–15, 1977. CBMS Regional Conference Series in Mathematics, vol. 40
     (American Mathematical Society, Providence, 1979)
112. A.D. Myshkis, Generalizations of the theorem on a fixed point of a dynamical system inside
     of a closed trajectory. Mat. Sb. **34**(3), 525–540 (1954) (in Russian)
113. L. Nirengerg, in *Topics in Nonlinear Functional Analysis.* Revised Reprint of the 1974
     Original. Courant Lecture Notes in Mathematics, vol. 6, New York University, Courant
     Institute of Mathematical Sciences, New York (American Mathematical Society, Providence,
     2001)
114. V.V. Obukhovskii, On some fixed point principles for multivalued condensing operators.
     Trudy Mat. Fac. Voronezh Univ. **4**, 70–79 (1971) (in Russian)
115. V. Obukhovskii, N.V. Loi, S. Kornev, Existence and global bifurcation of solutions for a class
     of operator-differential inclusions. Differ. Equat. Dyn. Syst. **20**(3), 285–300 (2012)
116. V. Obukhovskii, P. Zecca, V. Zvyagin, On coincidence index for multivalued perturbations of
     nonlinear Fredholm maps and some applications. Abstr. Appl. Anal. **7**(6), 295–322 (2002)
117. V. Obukhovskii, P. Zecca, V. Zvyagin, An oriented coincidence index for nonlinear Fredholm
     inclusions with nonconvex-valued perturbations. Abstr. Appl. Anal. Art. ID 51794, 21
     p. (2006)
118. V. Obukhovskii, P. Zecca and V. Zvyagin, On some generalizations of the Landesman-Laser
     theorem. Fixed Point Theory **8**(1), 69–85 (2007).
119. H. Okochi, On the existence of anti-periodic solutions to a nonlinear evolution equation
     associated with odd subdifferential operators. J. Funct. Anal. **91**(2), 246–258 (1990)
120. A.I. Perov, V.K. Evchenko, *Method of Guiding Functions* (Izdat.-Poligr. Centr Voronezh Gos.
     Univ., Voronezh, 2012) (in Russian)
121. W.V. Petryshyn, On the approximation-solvable of equations involving A-proper and pseudo
     A-proper mappings. Bull. Am. Math. Soc. **81**, 223–312 (1975)
122. S. Pinsky, U. Trittmann, Anti-periodic boundary conditions in supersymmetric discrete light
     cone quantization. Phys. Rev. D **62**, 087701 (2000)
123. T. Pruszko, A coincidence degree for L-compact convex-valued mappings and its application
     to the Picard problem of orientors fields. Bull. Acad. Polon. Sci. Sér. Sci. Math. **27**(11–12),
     895–902 (1979/1981)
124. P. Rabinowitz, Some global results for nonlinear eigenvalue problems. J. Funct. Anal. **7**,
     487–513 (1971)
125. D. Rachinskii, Multivalent guiding functions in forced oscillation problems. Nonlinear Anal.
     26(3), 631–639 (1996)
126. L. Schwartz, *Cours d'Analyse.* 1, 2nd edn. (Hermann, Paris, 1981)
127. J. Shao, Anti-periodic solutions for shunting inhibitory cellular neural networks with time-
     varying delays. Phys. Lett. A **372**(30), 5011–5016 (2008)
128. E. Tarafdar, S.K. Teo, On the existence of solutions of the equation $Lx \in Nx$ and a
     coincidence degree theory. J. Aust. Math. Soc. Ser. A **28**(2), 139–173 (1979)
129. M. Väth, New beams of global bifurcation points for a reaction-diffusion system with
     inequalities or inclusions. J. Differ. Equat. **247**(11), 3040–3069 (2009)
130. M. Väth, in *Topological Analysis: From the Basics to the Triple Degree for Nonlinear
     Fredholm Inclusions.* De Gruyter Series in Nonlinear Analysis and Applications, vol. 16
     (Walter de Gruyter, Berlin, 2012)
131. J.R. Webb, S.C. Welsh, in *A-Proper Maps and Bifurcation Theory.* Lecture Notes in
     Mathematics, vol. 1151 (Springer, Berlin, 1985), pp. 342–349
132. S. Wang, On orientability and degree of Fredholm maps. Mich. Math. J. **53**, 419–428 (2005)
133. T. Yoshizawa, in *Stability Theory by Liapunov's Second Method.* Publications of the
     Mathematical Society of Japan, vol. 9 (The Mathematical Society of Japan, Tokyo, 1966)
134. P. Zecca, V.G. Zvyagin, V.V. Obukhovskii, On the oriented coincidence index for nonlinear
     Fredholm inclusions. Dokl. Akad. Nauk **406**(4), 443–446 (2006) (Russian). English transla-
     tion: [J] Dokl. Math. **73**(1), 63–66 (2006)

135. V.G. Zvyagin, The existence of a continuous branch for the eigenfunctions of a nonlinear elliptic boundary value problem. Differencial'nye Uravnenija **13**(8), 1524–1527 (1977) (in Russian)

136. V.G. Zvyagin, The oriented degree of a class of perturbations of Fredholm mappings and the bifurcation of the solutions of a nonlinear boundary value problem with noncompact perturbations. Mat. Sb. **182**(12), 1740–1768 (1991) (Russian); English translation: Math. USSR-Sb. **74**(2), 487–512 (1993)

137. V.G. Zvyagin, N.M. Ratiner, Oriented degree of Fredholm maps of nonnegative index and its application to global bifurcation of solutions, in *Global Analysis – Studies and Applications, V*. Lecture Notes in Mathematics, vol. 1520 (Springer, Berlin, 1992), pp. 111–137

Index

ANR-space, 11
$B_{C_T}(0, r)$, 24
$B_C(0, r)$, 24
$B_E(0, r)$, 24
$C([0, T]; E)$, 24
$CG(\tilde{E})$, 132
$CJ(X, Y)$, 12
C_{00}, 23
$C_T([0, T]; E)$, 24
$C_{pr}^1[0, 1]$, 145
$Coin(L, \mathscr{F})$, 19
$Fix\mathscr{F}$, 12
J–multimap, 11
$J(X, Y)$, 11
$J^c(X, Y)$, 43
L-homotopic, 20
$L^p([0, T]; E)$, 24
R_δ-set, 10
T-non-recurrence point, 28
$W^{k,p}([0, T]; E)$, 24
$W_T^{k,p}([0, T]; E)$, 24
$(f \in \Phi_k C^1(Y))$, 131
ε-approximation, 8
    regular, 9
$\mathscr{BC}(H)$, 83
$\mathscr{BC}(Y)$, 89
$\mathscr{S}$, set of all non-trivial solutions, 52
$Cv(Y)$, 3
$Kv(Y)$, 3
$Pv(Y)$, 3

$CG^+(\tilde{E})$, 132

absolute retract, 11
anti-periodic problem, 49

anti-periodic solutions, 49
approximation solvable, 70
aspheric, 10

bifurcation index, 51
bifurcation point, 51
Bohnenblust-Karlin theorem, 17

$C(Y)$, 3
Castaing representation, 5
Clarke's generalized gradient, 37
coercivity condition, 32
coincidence degree, 20
coincidence index
    oriented, 134
coincidence point, 19
coincidence points set, 133
contractible, 10

decomposition, 12
Differential game, 45
    finite, 45
    infinite, 45
    of pursuit, 45

embedding, 10
evaluation map, 26

feedback control system, 102
Filippov implicit function lemma, 103
fixed point, 12
fixed point principle, 16

Fredholm
    map of index k, 131
    oriented structure, 132
    atlas, 132
    linear operator of zero index, 19
    oriented atlas, 132
    oriented structure, 132
    structure, 132
function
    regular, 37
fundamental subset, 140

generalized derivative, 36
generalized gradient, 84
generalized periodic problem, 44
global bifurcation index, 157
gradient functional differential inclusion, 42
Gronwall Lemma, 27
guiding function, 33
    integral, 72
    local, 57
    local integral, 59
    non-smooth, 38
    non-smooth integral, 40
    strict, 32
    strict non-smooth, 38
    weak, 35

Hausdoff metric, 3
homotopy
    Additive dependence on the
        domain, 16
    invariance property, 15

inclusion
    one-parameter family, 51
    operator, 71
index
    of V, 38
    of a non-degenerate potential, 32
    local of a non degenerate potential, 57
    of a nondegenerate projectively
        homogenous potential, 72
    of the map $V^{\sharp}$, 95
    of the non-degenerate projectively
        homogeneous potential, 85
infinite delay, 64
Ize's lemma, 52

$K(Y)$, 3

Leray-Schauder boundary
    condition, 18
locally contractible, 10
locally Lipschitz, 36
Lusin property, 6

measurable selection, 5
measure of noncompactness, 140
MNC, 140
    Hausdorff, 140
    Kurathowski, 140
    monotone, 140
    nonsingular, 140
    real, 140
multifield, 12
    family, 12
    topological degree, 14
multifunction, 5
    integrable, 6
    measurable, 5
    step multifunction, 6
    strongly measurable, 6
multimap, 1
    $CJ(X, Y)$, 12
    L-compact, 19
    L^{p}-upper Carathéodory, 7
    approximable, 11
    closed, 2
    compact, 3
    complete pre-image, 2
    completely u.s.c., 4
    continuous, 2
    convex closure, 5
    feedback, 102
    finite-dimensional
        approximation, 14
    homotopic, 15
    image, 1
    l.s.c., 2
    locally compact, 3
    lower semicontinuous, 2
    quasicompact, 3
    small preimage, 1
    topological degree, 13
    u.s.c., 2
    upper Carathéodory, 7
    upper semicontnuous, 2
multioperator, 26
multivalued map (see multimap), 1
multivalued vector field (see multifield), 12

neighborhood retract, 10

oddfield, 18
oriented coincidence index
    Additive dependence on the domain
        property, 138
    coincidence point property, 136
    homotopy invariance property, 137
    of a β-condensing triplet, 141
    of a β-condensing triplet
        Coincidence Point Property, 143
        homotopy invariance property, 142
    of a compact triplet, 138

partial differential equation
    control problem, 80
phase space $\mathscr{B}$, 22
potential
    direct, 37
    local non-degenerate, 56, 94
    non-degenerate, 31
    non-degenerate non-smooth, 37
    projectively homogeneous, 71, 84
principle of forbidden direction, 17
principle of map restriction, 16
projection, 69
proper, 132

retract, 10

Schauder projection, 14
Schroedinger equation, 118
second order differential
        inclusiion, 105
singular point, 12
solution
    T-periodic of inclusion, 26
    local of inclusion, 25
    of inclusion, 25
    global of inclusion, 25
    trivial, 53
subdifferential, 37
sublinear growth, 8
superposition multioperator, 8
superpositionally measurable, 111

translation multioperator, 26
triplet
    β-condensing, 140
    compact, 133
    compact approximation, 141
    finite dimensional, 134

u.s.c.
    weakly, 80

LECTURE NOTES IN MATHEMATICS

 **Springer**

Edited by J.-M. Morel, B. Teissier; P.K. Maini

Editorial Policy (for the publication of monographs)

1. Lecture Notes aim to report new developments in all areas of mathematics and their applications - quickly, informally and at a high level. Mathematical texts analysing new developments in modelling and numerical simulation are welcome.

   Monograph manuscripts should be reasonably self-contained and rounded off. Thus they may, and often will, present not only results of the author but also related work by other people. They may be based on specialised lecture courses. Furthermore, the manuscripts should provide sufficient motivation, examples and applications. This clearly distinguishes Lecture Notes from journal articles or technical reports which normally are very concise. Articles intended for a journal but too long to be accepted by most journals, usually do not have this "lecture notes" character. For similar reasons it is unusual for doctoral theses to be accepted for the Lecture Notes series, though habilitation theses may be appropriate.

2. Manuscripts should be submitted either online at www.editorialmanager.com/lnm to Springer's mathematics editorial in Heidelberg, or to one of the series editors. In general, manuscripts will be sent out to 2 external referees for evaluation. If a decision cannot yet be reached on the basis of the first 2 reports, further referees may be contacted: The author will be informed of this. A final decision to publish can be made only on the basis of the complete manuscript, however a refereeing process leading to a preliminary decision can be based on a pre-final or incomplete manuscript. The strict minimum amount of material that will be considered should include a detailed outline describing the planned contents of each chapter, a bibliography and several sample chapters.

   Authors should be aware that incomplete or insufficiently close to final manuscripts almost always result in longer refereeing times and nevertheless unclear referees' recommendations, making further refereeing of a final draft necessary.

   Authors should also be aware that parallel submission of their manuscript to another publisher while under consideration for LNM will in general lead to immediate rejection.

3. Manuscripts should in general be submitted in English. Final manuscripts should contain at least 100 pages of mathematical text and should always include

   - a table of contents;
   - an informative introduction, with adequate motivation and perhaps some historical remarks: it should be accessible to a reader not intimately familiar with the topic treated;
   - a subject index: as a rule this is genuinely helpful for the reader.

   For evaluation purposes, manuscripts may be submitted in print or electronic form (print form is still preferred by most referees), in the latter case preferably as pdf- or zipped psfiles. Lecture Notes volumes are, as a rule, printed digitally from the authors' files. To ensure best results, authors are asked to use the LaTeX2e style files available from Springer's web-server at:

   ftp://ftp.springer.de/pub/tex/latex/svmonot1/ (for monographs) and
   ftp://ftp.springer.de/pub/tex/latex/svmultt1/ (for summer schools/tutorials).

Additional technical instructions, if necessary, are available on request from lnm@springer.com.

4. Careful preparation of the manuscripts will help keep production time short besides ensuring satisfactory appearance of the finished book in print and online. After acceptance of the manuscript authors will be asked to prepare the final LaTeX source files and also the corresponding dvi-, pdf- or zipped ps-file. The LaTeX source files are essential for producing the full-text online version of the book (see http://www.springerlink.com/openurl.asp?genre=journal&issn=0075-8434 for the existing online volumes of LNM). The actual production of a Lecture Notes volume takes approximately 12 weeks.

5. Authors receive a total of 50 free copies of their volume, but no royalties. They are entitled to a discount of 33.3 % on the price of Springer books purchased for their personal use, if ordering directly from Springer.

6. Commitment to publish is made by letter of intent rather than by signing a formal contract. Springer-Verlag secures the copyright for each volume. Authors are free to reuse material contained in their LNM volumes in later publications: a brief written (or e-mail) request for formal permission is sufficient.

Addresses:
Professor J.-M. Morel, CMLA,
École Normale Supérieure de Cachan,
61 Avenue du Président Wilson, 94235 Cachan Cedex, France
E-mail: morel@cmla.ens-cachan.fr

Professor B. Teissier, Institut Mathématique de Jussieu,
UMR 7586 du CNRS, Équipe "Géométrie et Dynamique",
175 rue du Chevaleret
75013 Paris, France
E-mail: teissier@math.jussieu.fr

For the "Mathematical Biosciences Subseries" of LNM:

Professor P. K. Maini, Center for Mathematical Biology,
Mathematical Institute, 24-29 St Giles,
Oxford OX1 3LP, UK
E-mail : maini@maths.ox.ac.uk

Springer, Mathematics Editorial, Tiergartenstr. 17,
69121 Heidelberg, Germany,
Tel.: +49 (6221) 4876-8259

Fax: +49 (6221) 4876-8259
E-mail: lnm@springer.com